P9-BTL-919

Date Due

Ancestral Books in the Management of Organizations

A 31-volume facsimile series
reproducing classic works in the field.

Edited by
Arthur P. Brief
Graduate School of Business Administration
New York University

A Garland Series

The Changing Culture
of a Factory

Elliot Jaques

Garland Publishing, Inc.
New York • London
1987

For a complete list of the titles in this series
see the final pages of this volume

This facsimile has been made from a copy
in the Yale University Library.

First published in 1951 by
Tavistock Publications Limited
in collaboration with Routledge and Kegan Paul Limited.
Reprinted by permission of the Tavistock Institute.

Library of Congress Cataloging-in-Publication Data

Jaques, Elliot.
The changing culture of a factory.

(Continuity in administrative science)
Reprint. Originally published: London : Tavistock, 1951.
Bibliography: p.
Includes index.
1. Metal-workers—Great Britain.
2. Metal work—Great Britain—Personnel management.
3. Glacier Metal Company—Employees. I. Title. II. Series.
HD8039.M5J37   1987   671'.068   86-25815
ISBN 0-8240-8212-5   (alk. paper)

The volumes in this series are printed on
acid-free, 250-year-life paper.

Printed in the United States of America

THE CHANGING CULTURE
OF A FACTORY

THE
CHANGING CULTURE
OF A FACTORY

ELLIOTT JAQUES,
M.D., Ph.D.

TAVISTOCK PUBLICATIONS LTD.

First published 1951
by Tavistock Publications Limited
in collaboration with
Routledge and Kegan Paul Limited
Broadway House, 68–74 Carter Lane, London, E.C.4,
and printed in Great Britain
by Butler & Tanner Limited, Frome and London

Second impression 1952
Third impression 1957
Fourth impression 1961

RESEARCH TEAM MEMBERS

Elliott Jaques, M.D., Ph.D., Director of the Study.

A. Mitchell, M.A.	E. L. Trist, O.B.E., M.A.
A. K. Rice, B.A.	G. L. Ladhams (part-time consultant)

W. Annandale, B.A.	K. W. Bamforth, B.Comm.
J. M. M. Hill, B.A.	Miss R. Hurstfield, B.Sc.
Miss R. Carey, B.Sc.	I. Leff, S.B.
L. Morrell	T. B. Ward, M.B.E.

▼

ACKNOWLEDGEMENTS

ALTHOUGH I appear as author of this book, its preparation has been the result of the closest collaboration between myself, my colleagues from the Tavistock Institute who have worked with me on the project, and members of the Glacier Metal Company. Through the Managing Director, Mr. W. B. D. Brown, and his Works Council and management colleagues, I should like to acknowledge the co-operation of the members of the firm—and in particular those (far too numerous to mention individually) who have actively participated in the work and have read and criticized those parts of the manuscript which cover events with which they were directly concerned or about which they had special knowledge.

More personal acknowledgement can be made of a smaller number of members of the firm who took part in the work of three committees of the Works Council. The first were members of the Steering Committee, Dr. G. H. Gange (Chairman of the Council,) T. C. Challinger, D. J. Clarkson, and W. H. Morton. The second were members of the Project Subcommittee which later became the Internal Development Committee, and who took a leading part in the planning and development of the project— D. J. Clarkson, R. B. Finch and P. D. Liddiard (Chairmen), Miss R. Healey, and G. W. Cannon, J. Dalziel, J. Duffy, L. Emsley, R. Hartley and T. Hungerford. Third were members of the Public Relations Committee of the Works Council who did such careful work in the final appraisal and criticism of the manuscript as a whole: N. E. Hughes and L. Saunders (Chairmen), W. B. D. Brown, T. S. Burnside, J. Thompson and W. E. Wain.

I have received invaluable help from Sir George Schuster, Mrs. Margaret Cole, Dr. A. T. M. Wilson, and J. Harvard-Watts, who have read either the whole or parts of the manuscript. My special thanks are due to Eric Trist for extensive criticism of the manuscript at each stage of its growth, and for stimulating and helpful discussion of the material.

Acknowledgements

The book has been accepted by the Department of Social Relations at Harvard as a Ph.D. thesis. I should like to acknowledge my debt to members of the department for the flexibility in outlook and the breadth of interest which allowed them to make this possible.

Many of the diagrams were prepared by W. Annandale. Miss Jean Allison, Mrs. Wilma Atkins, Mrs. Francesca Bion and Mrs. Enid Leff performed valuable service as Research Assistants. Mrs. Elizabeth Smith typed countless versions of the manuscript, ably assisted by Miss Margaret Hipkin and Miss Doreen Paparritor.

E. J.

2 Beaumont Street,
   London, W.1.   December, 1950.

CONTENTS

Contents

DIAGRAMS

TABLES

INTRODUCTION

DURING the past fifty years there have been increasing opportunities to test the view that a scientific approach may assist the understanding of social problems and provide one means of improving decisions taken in tackling them. One recent opportunity has arisen through the interaction of three factors in the British industrial field—economic pressure for maximum productivity; severe limitation of re-equipment or expansion; and the altered social climate of a full employment situation. In these circumstances the part played in productivity by social and psychological factors has become particularly obvious, and increased support has been forthcoming for research.

The project described in the present book emerged from this background. It shows how the large-scale problems of British industry are reflected in the social development of a single industrial unit—a light engineering firm of some fifteen hundred people. The attempt of this firm to understand its problems is the theme of the book; but there are also illustrations of the close relationship between technological and social development, and of the difficulties in applying scientific method outside the laboratory in the world of social reality. This last point is a main interest of the Tavistock Institute of Human Relations, the research organization which collaborated in the work reported; and it may therefore be appropriate to offer some remarks on the Institute's general methods of social research, as represented in this industrial study.

The development of methods for systematic observation of human behaviour, and for understanding the psychological and social forces concerned in it, presents no more difficulty than another problem less frequently discussed—that of finding opportunities to observe, at first hand, those areas of human behaviour that are manifestly of importance in human life. One way of tackling this problem of access derives from views which become intelligible in terms of psychoanalytic theory. These views may be summarized in this way: first, that each one of us, in the course of development, has painfully worked out a set of assumptions as to what is real and what is important in determining our behaviour; secondly, that these assumptions give meaning to our lives and offer some protection from fear and uncertainty; and, thirdly, that even personal attempts to modify such deeply-rooted assumptions arouse anxiety and resistance which can only be overcome by serious psychological effort. Finally, under certain definable conditions—particularly those of professional work dealing with practical human and social problems —the need to overcome a difficulty, that is, the need for change, may offset anxiety and resistance to both access and change, and permit first-hand observation of important areas of behaviour.

There is considerable consistency in the conditions governing the emergence of professions concerned in such work, and of the contracts governing their activities. The limited power of adaptation of individuals and social organizations has led to the development of specialized functions and roles concerned with problems of individual or social change; and in some cases these have slowly become accepted as professions. In Institute projects, such as that to be described, a comparable situation may be said to exist: a professional person, who is also a research worker, co-operates with a client who has asked for assistance in dealing with problems for which he is responsible. A relationship of this kind is perhaps most obvious in the clinical research of medicine; but explicit or implicit professional responsibility is likely to be an important condition for most studies of forces affecting individual and group behaviour, and particularly for studies of actual social development.

In collaborative attempts to understand problems of this kind, the Institute uses a field theory approach, and regards behaviour as determined by a field of interrelated forces. The task of the research worker is seen as that of discovering the existence and nature of these different forces and showing their relationships in facilitating or inhibiting social development. The study of social development implies observation over a period of time. It will be noted that in the present study continuous work over nearly three years was needed before the first major changes were accomplished and recorded. These changes occurred within a single industrial firm, and thus no definite conclusions can be drawn as to problems of social change in other situations. The recorded observations, however, support the view that the more precisely the pattern of forces determining social behaviour can be traced out in relation to any one particular problem, the easier it will be for the results to be fitted into other observations.

On this point the reader will be able to judge for himself how far the descriptions given appear relevant to problems with which he is directly familiar.

The Institute's position with regard to the significance of measurement and quantitative treatment should be indicated. Opportunity for first hand observation may make it possible to identify and define the main forces concerned with social behaviour, and this is regarded as a first, and necessary, step in quantification. Success in obtaining a general picture of these forces leads on to the use of concepts of measure appropriate to a field theory—for example: magnitude, distance, direction, strength, and valence. In the particular study here reported it has been possible to identify and define many of the forces affecting social behaviour and social development in specific industrial situations. If research interests alone had been relevant, measurement techniques could have been introduced as soon as description of the field of forces, at a first level of approximation, had been achieved. Within the approach described, however, any action for research purposes alone, without regard to the needs of the client, would be regarded as a breach of the professional role of the research worker. This limitation is in part offset by the common experience that attempts to use detailed measuring instruments, before obtaining adequate description of the forces concerned in a particular type of behaviour, are apt to lead to irresolvable problems of assessing the significance of obtained measures. Techniques of measurement need to be appropriate to each stage and level of description of such forces. It is anticipated that in follow-up studies now being undertaken at the request of various sections of the firm, opportunities will arise for more advanced steps in measurement; members of the firm now require data of this kind for their own purposes, as a result of the work reported.

In scientific work of all kinds there has been a growing realization of the part played by the observer himself as one factor determining his observations. Scientific research in relatively new fields of work shows the importance of the observer describing in detail his role, his methods, and his view of the character and limits of his field of observation. If this is accepted, a general question of both practical and theoretical significance must now be raised in relation to the approach exemplified in this study: how far may research workers employing such an approach, in whatever field, unwittingly overlook important factors because of the limits set by that approach and by their concern with a particular set of hypotheses? A partial answer to this question is given in the body of this report, since it has played a prominent part in the work. A professional relationship between a research worker and a client implies shared responsibility for what happens as a result of the joint analysis of the problem which brought them together. The action taken in tackling the problem, and the results of this action, are therefore open to assessment, not only by the research worker and by those with whom he is directly collaborating, but also by others who are affected by what has, or has not, been achieved. Where action based on an analysis of a particular problem fails, the research worker is compelled by his role to accept the incompleteness both of his observations and his analysis, and to discover the shortcomings of his approach. In this way there is by necessity some testing of his hypotheses, and some assessment of their adequacy for that particular problem. Practical success in employing a hypothesis represents no more than a stage in theoretical development, and does not remove the need for constantly subjecting it to the test of further application and further critical scrutiny. This need is a general characteristic of professional and scientific work of all kinds; but it demands special attention wherever the research worker exposes himself to the full complexity of real life situations.

The text of this report has been subjected to rigorous criticism both by the members of the firm directly participating in the different parts of the work described, and by the Works Council, which has been responsible for the project on behalf of the firm. In this way there has been provided some check of the reliability of the observations, hypotheses, and actions recorded. There is a further point to be noted in this connection: when a particular difficulty has been tackled and resolved, those involved have a sense of having completed a specific task. This sense of completion (even though solution of the problem may represent no more than part of a larger task) removes many of the objections to the description and publication of the experiences and results; for, in a sense, these have become past history, and may be made public without embarrassment.

The value of research undertaken by the method described will only become clear as the results of its use become available for critical assessment, but, within the limits defined, it may be said to provide opportunities of exploring at first hand a number of significant areas of social behaviour, and in particular the dynamics of social groups. The specific study here reported shows how unconscious forces in group behaviour, and the unwitting collusion between groups for purposes of which they are only dimly aware, are important factors in the process of social adaptation. It also shows—and this is perhaps more important— how such forces, and such unrecognized collusion, may be a main source of difficulty in implementing agreed plans for social development and social change. In this connection, practical interest attaches to the description given of methods of the working-through of problems, so that those concerned may learn how to take account of a wider span of factors in making decisions, and hence achieve greater flexibility of performance. In social science terms, the work illustrates the interplay of social structure (defined in role systems), of culture, and of personality, although in this study these concepts have not been used to develop a general theory of social behaviour. They are used for the specific purpose of understanding field observations and experiences. Such integrative analyses of concrete situations are complementary to purely theoretical development, and can in this way hope to contribute to the growth of a basic social science.

Such matters may at first sight appear a little distant from the life and work of industry, but those concerned with industrial affairs may find in the work reported some indications of the nature of problems which they themselves are increasingly called upon to tackle. The relation between executive and consultative procedures, and the roots of such problems in the sanctioning of authority, provide one example of this; but there are also significant reminders of the inevitable interaction between technological and social development, and of the need to consider these two aspects of industrial life simultaneously rather than consecutively.

Tavistock Institute of Human Relations,
2 Beaumont Street,
London, W.1.

A. T. M. Wilson, M.D.,
Chairman,
Management Committee.

PART ONE

BACKGROUND DATA

The purpose of the project, and certain aspects of the methods and approach employed, will be outlined and illustrated by the events of the first three months, when the project got under way. A broad background of the firm is then given by means of a description of its organization in 1948 and a rapid historical sketch showing how this organization grew.

CHAPTER ONE

THE CONCEPTION AND INITIATION OF THE PROJECT

THIS book is a case study of developments in the social life of one industrial community between April, 1948, and November, 1950. It is a progress report, and gives an account of research participation in attempting to deal with the day-to-day problems experienced by the factory in its efforts to find a more satisfying working life consistent with the demands of a competitive industrial situation. As a case study, the book is not intended to be a statement of precise and definite conclusions. It is written in three sections: the first of these presents introductory, background, and historical material; the second consists of five independent case studies of problems experienced in five separate parts of the factory; the third and final section draws some of the material together, as well as presenting certain limited conclusions arising directly from the case studies.

The factory has a wide reputation for its social policies and is regarded by many as a special case: it has introduced most of the modern methods in progressive management and has made certain innovations which are very much its own. The quality of group relations has steadily improved over the past ten years, and the firm is now accepted generally by its members at all levels as a very satisfactory place in which to work. As regards the changing pattern of social stresses and strains which are experienced, some of these arise from the fact that the advanced development of social relations has thrown up new and unfamiliar problems, but many of the difficulties are similar to those which occur in other communities—whether industrial or educational, family or neighbourhood, urban or rural. The capacity of the firm to deal with difficulties, as and when they occur, is one indication of its vitality. Because of this it has been possible to collaborate with the personnel of the factory in exploring,

3

illuminating and recording many of the underlying aspects of these stresses—undercurrents which in ordinary situations are not accessible either to the community or to the research worker, since the level of security is insufficient to allow either their expression or their recognition.

The material presented is a first report of a long-term project, whose purpose is to study some of these less accessible sources of group stress. It is part of a more general programme of research, administered through the Medical Research Council and approved by the Human Factors Panel of the Committee on Industrial Productivity which was set up in 1948 by the Lord President of the Council.[1] The Human Factors Panel was constituted a body composed of nine members—three Government representatives, a delegate from the British Employers' Confederation, a delegate from the Trades Union Congress, an independent member, and three representatives from independent social science research institutions. The views of the Chairman on industrial relations commanded wide respect among both management and workers in Glacier. The delegate of the Trades Union Congress was the National President of the Amalgamated Engineering Union, one of the main unions in the factory. Sponsorship by a panel with these two particular members, together with the accepted independence of the Medical Research Council, were influences of considerable importance in sanctioning the beginning of the work.

THE OPERATIONAL PLAN OF THE PROJECT

The plan to be followed in this book reflects the operational plan of the project itself, beginning with a general study of the factory and the forces affecting it as a total community, and moving on to more detailed descriptions of events occurring in component sections. Such a plan derives from field theory, which, briefly stated, holds that a particular event may best be understood as the outcome of other interacting events in the larger field in which it occurs; that is to say it is unlikely that notions of simple cause and effect will explain social behaviour. For example, to describe the effects of introducing a piece-rate system or a scheme of joint consultation into a factory, it would

[1] No responsibility attaches to either of these official bodies for material used or statements made in this book.

4

be necessary, in terms of this approach, to take into account the setting in which these events occur at the time, such as the general morale situation in the particular factory concerned, the rates of pay, the structure and nature of working groups, and the quality of supervision, and, equally, the larger social forces emanating from the general economic situation, the competitive position of the factory, and the characteristics of the local community. Not that in what follows such a complete background will be presented. But the pattern has been to move, as far as was possible, from the more general to the more specific forces, which can then be seen against the larger setting. Change and development being one of the essential characteristics of every social process, work of this kind is never completed, and the present project remains in this sense unfinished. The research worker must often rest content with the study of limited phases of processes which have gone on before his arrival and which will continue after his departure.

The Complexity of Social Change

Another way in which the plan of the book reflects the growth of the project is that, while the early parts deal with material deriving from a background study initiated by the Research Team, the later portions are concerned with problems brought to the Research Team by various sections of the firm. This shift took place as the Research Team acquired greater acceptability as an independent group with professional status; an extremely important process, which has continued, though at a slow pace and with uneven development, in different parts of the factory.

About each of the problems for which assistance was sought, the Research Team made three assumptions: that the particular problems complained of were unlikely to be the only—or indeed the main—sources of difficulty; that no simple causes, or solutions, would be found; and that although assistance was sought, resistances to change were likely to occur as the real situation was further explored. Such assumptions, while common in some kinds of medical work, are nevertheless not universally accepted as a basis for the study of human behaviour. Frequently, simplified answers are sought—panaceas which will cure a variety of difficulties with a minimum of discomfort—with the research worker too often conforming, or being forced to conform, to the wishes rather than meeting the needs of those presenting urgent

5

demands under the stress of industrial practice. The introduction of piece-rates, or of better personnel management, the use of joint consultation, profit sharing, or of improved works information—the list is long of remedies, which, in spite of their many advantages, have on the whole led to disappointment whenever those concerned have been able and willing to face or able to assess the total effects produced against the time, expense and effort put into what, after all, amount to no more than relatively simple changes of practice or procedure.

There is considerable evidence that social problems, on however small a scale, require for their solution much more than administrative or technical changes which still leave intact the underlying system of values and the familiar attitudes and outlook which form the culture or way of life of the community. The Hawthorne experiment, the Poston study, the Yankee City surveys[1] all bear witness to the need to approach human problems with due deference to the complexity of the factors at work, and for the unique manner in which these factors operate and interweave in any situation at a given point in time. It was assumed in this study that the simplest approach was to accept the complexity of social reality, and the most rational, to accept the irrationality of many of the unrecognized forces which contribute to social behaviour.

The Design of the Project

The design chosen for the project was to collaborate with one firm which would be willing and interested to study and develop its methods for creating satisfactory group relations; it being felt that an advanced firm would provide the most suitable field of study. For such a firm would have already experienced the limited return to be realized from the introduction of merely procedural changes and partial remedies of the types discussed in the previous section. On this account it was likely that such a firm would already be feeling in itself considerable need to confront and understand some of the deeper forces with which the research was concerned. It would tend, therefore, to welcome an approach for collaboration with a technical group engaged on the study of these forces from the point of view of social science

[1] Descriptions of these studies will be found in: Roethlisberger and Dickson, *Management and the Worker*, Harvard University Press, *1947*; Leighton, *The Governing of Men*, Princeton University Press, *1945*; Warner and Lunt, *The Social System of a Modern Factory*, Yankee City Series 4, Yale University Press, *1947*.

matters, and be capable of tolerating the strain and disturbance inevitably associated with research and pioneer developments, especially in a field such as group relations.

It was anticipated that such work with one firm, and that a firm which might in some respects be employing uncommon practices, could produce results of general interest; for however diverse the character of problems of industrial relations may appear on the surface, in many ways these problems will have common underlying features, whatever the industry. If this assumption holds true, a study which contributes to the teasing out and illumination of some of the psychological and social roots of stresses in group relations in one factory, will contribute to the illumination of similar problems elsewhere.

The Glacier Metal Company was approached, both because it was regarded by the Institute as fulfilling many of the general conditions necessary for the projected type of collaborative field study, and because in intermittent working contact with the firm for a period of nearly two years the Institute had found it an organization that provided specially favourable conditions for following out the principle that management, supervision, and workers each should independently agree to collaborate in any work undertaken. On the other hand there were certain difficulties about Glacier which were recognized. It was not fully organized as a union shop, nor was it a member of the Engineering and Allied Employers' Federation. Both these features were felt in part to arise from the very fact that the factory was experimenting with advanced and novel methods. It presented an uncommon opportunity to investigate how far a firm which had developed social practices of a more advanced character than was yet usual in British industry could nevertheless maintain satisfactory connections with the larger social units of which it forms a part both on the employer and the worker side.

THE FIRST THREE MONTHS

Field study began formally in July, 1948, but the Institute had first approached the firm in April of that year, the intervening three months being taken up with securing from all sections of the factory as full initial agreement to the undertaking of the project as it was possible to obtain. The Managing Director and his immediate subordinates were approached first. The purpose

7

of the project was described and answers given to such questions as were answerable. Although at this stage plans could only be outlined in vague and uncertain terms, he and they agreed to collaborate, in so far as it was possible for them to make any definite commitment themselves in face of the indeterminate nature of the Institute proposals.

This preliminary agreement from management was by no means automatically obtained, and might easily have been withheld had there not been some previous contact with the Institute. On the first of these occasions in 1946 the Managing Director had been most anxious to engage the collaboration of the Institute, despite the fact that his colleagues were opposed to it. The Institute had stated that it refused on principle any request for consultation unless not only the management group as a whole were agreed, but representatives of the workers also; such general agreement being regarded as an essential pre-condition of any technically effective and professionally responsible undertaking in this special line of work. The delay which this caused enabled the Institute to establish a first understanding with the firm of its seriousness over securing general agreement before collaboration could begin. It happened eventually that other members of the senior management group as well as the Managing Director desired the help of the Institute, and it was under these conditions that agreement, including that of supervision and of workers' representatives, was obtained for the first consultant arrangement undertaken between Glacier and the Institute. When collaboration in the present project was first proposed, several members of the Glacier management, feeling that the Managing Director was already investing too much of his time and interest in social experiments, and anxious lest production be affected adversely by still further involvement of this kind, expressed considerable reservation as to the wisdom of having research of this kind going on in the factory. As against this they showed interest in progressing towards good industrial relations; and this interest, along with the knowledge that the Managing Director was particularly set on having this project carried out in his factory, as well as their experience that the Institute would not take part in a project imposed from above, proved to be forces sufficiently strong to overcome the reservations that they had expressed.

The next step was for the matter to go to the Works Council.

Certain members of the management group now became anxious lest the workers' representatives should turn the whole scheme down in view of a general trade union antipathy towards, and scepticism about, industrial psychologists and psychiatrists. They suggested that the Chairman of the Human Factors Panel should be asked personally to come down to the Works Council meeting, since the high regard in which he was held throughout the factory would lend great weight in securing approval for the project, if he would say that an undertaking of this nature between Glacier and the Institute would be commendable to him. The Institute opposed this suggestion on the grounds that persuasion of this kind would run counter to the principle of getting genuine co-operation from the factory itself—a view which was shared by the Chairman of the Human Factors Panel himself. Accordingly the Institute representative went alone to the meeting of the Works Council scheduled for the end of April, where he was given an opportunity to describe the plan for the project, pointing out that although management had agreed to collaborate, independent agreement would be required also from the workers' side. To take care of this requirement, the Works Council referred the matter for independent consideration to the Works Committee, a body whose membership was entirely composed of elected representatives of the Glacier workers and whose next meeting took place two weeks later.

The Workers Agree to Collaborate

The Works Committee had twenty-four members; all were elected from their departments; all were trade unionists, and most shop stewards. This first meeting between the Works Committee and the Institute representative lasted five hours, and took place during a fierce thunderstorm which passed virtually unnoticed. Some members evinced a not unfriendly interest in the social sciences and industrial psychology; all wanted to know in detail what collaboration on the project would mean and entail.

"Who is behind this work?" The purpose of the Committee on Industrial Productivity and the scope of the programme of the Human Factors Panel were explained.

"Where do the Trade Unions stand, and what do they feel about the research?" Details were given about the official representation of the General Council of the Trades Union Congress on the Panel which was sponsoring the research.

9

"Who is paying for it, and what will it cost the firm?" The size of the grant received by the Institute and the kind of staff supported by such a grant were revealed, and an explanation was offered of the reasons for its administration by the Medical Research Council.

All these were straightforward questions, to which straightforward answers could be given. More difficult inquiries were to follow. "How long will the research last?" "How do you propose to go about it?" "What specifically do you intend to do?" The Institute could say that it had asked for, and had secured, an initial grant for one year, but did not know how long the research might actually last. Dilemmas arising from the vagueness and unpredictable course of the project were presented as real problems which the factory would be required to share along with the Research Team who hoped to develop new and more specific techniques as work got under way.

One of the main reasons for this questioning was that despite the fact that his Research Team would include between them a considerable range of social science and medical skills and that he himself was a doctor, the Institute representative nevertheless appeared solely as an industrial psychologist, and industrial psychology was not the most popular term among the workers. To them it smacked of efficiency engineering and speeding up and rate cutting, and at best was held suspect as a technique used by the management, which only indirectly or inadvertently, if at all, could be of benefit to the workers.

The questions clearly demanded more answering, therefore, and so the Institute's wish not to work with or for either the management or the workers, or any other group, but to collaborate with the factory as a whole by being responsible to the Works Council, was explained in some detail. There was no specific blue-print for the work. The most appropriate topics of study, particularly those events whose study might at one and the same time be of value to the firm and to industry generally, would be chosen in collaboration with the Works Council as the work progressed.

Such a vague account must have seemed thin; but the fact that nothing was being hidden communicated itself, and the workers decided to collaborate, subject to sanction from their trade union officials and agreement from the shop floor. Accordingly, at the request of the Works Committee, the Institute made contact with the representative of the Trade Union Congress, who had been one of the members of the Human Factors Panel responsible for sanctioning the project. Through him a letter stating trade union support for the project was sent from the Trades Union Congress to the national office of the Confederation of Shipbuilding and Engineering Unions, since the Confederation included within it all unions operating at Glacier. This step was reported to the Works Committee Secretary, who immediately passed on the information to his district union officers. After this nothing happened for a considerable time.

The initiative at this stage lay in the hands of the workers; but they began to feel anxious because they could get no "go ahead" signal from their unions—the London Regional Office of the Confederation having received no communication on the matter from the National Office. There was a period of delay lasting some weeks. But nothing could be done to hurry an answer for, having determined a policy of getting as full agreement as possible without pressure there was little to do but wait and hope. It was then discovered that there had been a breakdown in communication between the national and regional offices of the Confederation because of a series of conferences which had kept the national office staff away for many weeks. Social science research was a very low priority in comparison with the pressure of normal union business. However, a letter recommending collaboration in the project was subsequently sent through to the workers' representatives in the factory; and with this official sanction they in turn reported to the Works Council that they were in favour of having the project take place. They reported their recommendation to the Works Council on 18th July, and collaboration of the factory in the project was officially confirmed.

THE INDEPENDENT ROLE OF THE RESEARCH TEAM

The special care taken during these first three months, from the time of the first contact with the management until the management and the workers jointly decided to collaborate in the project, was to guard against the Research Team becoming involved in stresses between groups in the factory, and to establish from the beginning that a part of the approach used by them would be the avoidance of capture by any particular group.

Outsiders or newcomers to a community will, consciously or unconsciously, become to a certain extent a focus for the discharge of the community's own internal pressures. This is well known in everyday life. A new family, for example, moving into a neighbourhood where people are suspicious of each other, will often become a target for suspicion. It is as though the people in the neighbourhood maintain a semblance of harmony by denying their suspicion of each other and instead projecting such feelings outwards upon some convenient, commonly satisfactory object, in this case the newcomer. In such a manner, the social research worker is likely to have transferred on to him any unrecognized or unresolved stresses which individuals or groups in the community wish to avoid in relation to each other. In such a situation, the social research worker must discover for himself a role in which he belongs to no special section of the community; and one in which as far as possible he may have independent and equal relations with all. On scientific as well as ethical grounds, therefore, the building up of an independent relation to the factory as a whole was taken as the first project task, and the management and the workers, through their representatives, were separately offered the opportunity to collaborate in the research, or alternatively. to turn the whole project down.

The eventual adoption of the project by the Works Council after three months of consideration did not necessarily represent active interest on the part of the total factory, nor was it taken by the Research Team as having such a meaning. As far as the situation could be assessed, there were perhaps half a dozen members of the firm who positively and strongly wished the project to take place; there was also a larger group of thirty to fifty members who were moderately in favour, on the ground that some good might possibly result for the factory and for the country, and who were in such key positions as to allow them to decide on behalf of the total factory that a first trial period be undertaken; and finally there was the main body of members of the factory, who either had heard very little about the project, or were not particularly concerned. In short, there was very limited consultation between the members of top management and their subordinates, and between the workers' representatives and the shop floor, on whether the project should be undertaken. But, as will be seen later, difficulties in consultation because of the so-called apathy of the employees were one of the main

12

problems in the factory, and this became a central theme of study.

Defining the Independent Role

The Works Council, having determined to go ahead with the research, set up a Project Subcommittee of five to work with the Research Team in drawing up plans for the research. This subcommittee was composed of two management representatives —the Personnel Manager and the Medical Officer (who was also Chairman of the Works Council); and three workers' representatives—the Chairman of the Works Committee, the Secretary of the Works Committee, and a Works Committee member from the night shift.

During the first period of the project, through August and September, 1948, the Project Subcommittee and the Research Team met regularly at fortnightly intervals, to clarify the role of the Research Team, and to deal with such questions as who was to control the research, and, allied to it, was the Research Team to work for the firm. Out of these discussions there emerged a much clearer concept of the technical independence of the Research Team. It was accepted that Institute personnel could not be considered members of either the factory or the Project Subcommittee, but could collaborate with the factory through the Project Subcommittee with which body they would constitute a joint research committee.

A jointly agreed set of principles was drawn up covering the relations between the Research Team and the firm, and defining the general conditions under which the research would be carried out. These principles were set out in the following document which was posted on all notice boards in the factory.

(a) *The Research Team is responsible to and reports to the Works Council.*

(b) *The Works Council has delegated a Project Subcommittee which together with the Research Team will plan the programme and development of the Glacier Project.*

(c) *It is not the intention of the project that the Research Team should gather secret information from one person or group about another. The only material that will be of real value is information that is public and can be reported.*

(d) *Where any individual or group suggests a topic of study for the*

13

Research Team to look at, this shall only be done with the general approval of those likely to be affected by the results.

(e) All suggestions or requests will go to the Project Subcommittee which will consider them with the Research Team. A member or members of the Research Team with, as occasion demands, a member of the Project Subcommittee will then explore the matter.

(f) Any work done will be carried out under the following conditions:

> *(i) The Research Team to act only in an advisory or interpretive capacity. The team is not here to solve problems for Glacier. They may however be able to help with the continuing development of methods of getting a smoother organization.*

> *(ii) Nothing will be done behind anyone's back. No matter will be discussed unless representatives of the group are present or have agreed to the topic being raised.*

> *(iii) The Research Team will maintain professional confidence and will collaborate with those concerned in reporting back to the Project Subcommittee.*

These principles meant that the Research Team would work with any section of the factory at its request. In so doing, the members of the group concerned could talk about whatever they wished, but the Research Team for its part would comment only on relationships within the immediate group, and would most definitely not discuss any individuals or groups not present. In order to carry out these principles, the Research Team has limited its relationships with members of the factory to strictly formal contacts which have to do with project work publicly sanctioned by the Works Council. No personal relationships with Glacier members either inside or outside working hours have been entertained, except for such informal contacts as inevitably occur in walking through the works or eating in the various canteens, on which occasions the Research Team has tried to be scrupulously careful not to discuss the project, other than those aspects which had already been reported to the Works Council and, hence, were public property in the factory. This has meant refusing invitations to people's homes, to play tennis on the factory's courts, to discuss with individual members of a group occurrences which took place earlier at a meeting, and many other activities which could be construed as outside the terms of reference. After about a year the members of the factory learned that the Research Team was quite serious in its intention to adhere rigorously to the agreed terms, and all invitations of an informal character gradually came to an end.

This particular type of formal relationship was chosen for a number of reasons. It limited all contacts to those situations in which a particular section of the factory desired the assistance of the Research Team in tackling a specific problem or problems, and where the people involved were willing to use the formal and open channels to the Works Council to get such assistance. By this means the project work could be kept open and above board in the sense that everyone in the factory could know exactly what the Research Team was doing at all times, even though the team members would maintain professional confidence on the exact nature of the actual problems they might be dealing with in each group. Furthermore, the independence sought by the Research Team could not easily have been attained had personal friendships with members of the factory been tolerated during the course of the study. It is not consistent to profess an independent attitude towards a point of debate between two groups, and at the same time to have, outside of working hours, a social relationship with the members of one; and even if it were possible to maintain such social relationships and at the same time to avoid favouring consciously or unconsciously a particular section or sections of the factory, it would be impossible not to be regarded by the others as having been captured.

By thus maintaining an independent role in the factory, it was anticipated that assistance could be given most effectively to those groups which sought help with a problem. At the same time, if this proved correct, the Research Team was likely to be asked to help with increasingly difficult problems, and so brought into contact with areas of the life of the factory which in normal circumstances remain inaccessible except to those immediately involved. In this way it was hoped to overcome one of the most difficult tasks of social research—to gain access to data of importance which in ordinary circumstances might remain unrevealed.[1]

The model for this kind of research was taken from clinical

[1] Those interested in the technical aspects of the methodology of the present study will find a more general statement in "Interpretive Group Discussion as a method of facilitating Social Change", by the author, in *Human Relations*, Vol. I, No. 4, pp. 533–549.

medicine. In certain instances a patient may allow more treatment than is absolutely necessary, or perhaps new forms of treatment with a follow-up study of the effects, for possible benefits both to himself and to mankind in general. In this he co-operates with the doctor who accepts professional responsibility for the work done. This analogy to the doctor-patient relationship in a clinical research situation was not taken as applying exactly to a project in which a research team was working with a factory. There nevertheless seemed enough of value in the analogy to make it worth while to adopt the present approach in which an active working relation was sought with the factory on its day-to-day problems, and in which a fully professional role was accepted with its implications not only of confidentiality, but of responsibility for helping in working-through the problems raised.

POLICY GOVERNING PUBLICATION OF RESULTS

How to report the activities in real life of actual people in their *natural* community settings has always been a difficult problem for social scientists. Since the persons involved in the events in this book will be recognizable to anyone who knows the Glacier Metal Company, certain stringent conditions with regard to publication were undertaken from the beginning. One of the first matters agreed with the Works Council was a publications policy consistent with the desired collaborative relationship. This was posted in the factory in the following form:

> *Any public statements about the project by members of Glacier should be governed by the same policy as that covering any other public statement about the company. For their part, the Tavistock Research Team will make any public comments only in collaboration with Glacier, or after due consultation.*

This policy—made possible by the existence of a central representative body—was sought by the Research Team and not imposed by the firm, it being assumed that where the work of the project was successful in assisting the resolution of group problems there would be no difficulty in publication; and conversely, any differences arising over publication would indicate that the problems on which assistance had been sought were insufficiently worked-through, and, hence, were not ready to be written up. These undertakings have not meant that serious restrictions on publication have been in.posed by the factory in order to protect itself, or that the freedom of the investigators in making a scientific report has been impeded. Although real people emerge, with all the strengths and weaknesses characteristic of human beings, most of the problems described have been, or are now on the way to being, worked-through and resolved. Each section of the book was circulated to, and discussed in detail with, each of the groups particularly concerned, so that the task of writing up has been first and foremost a method of reporting back to the factory itself, the attendant discussions being regarded as an integral part of the work of the project. Following mutually agreed correction, modification, and elaboration between the Research Team and particular groups, each section has next been reported to Works Council. Finally, the report as a whole has been carefully checked and revised by that body, and reported to the factory generally.

Rather than hindering publication, the above procedure has ensured the active participation of those who know the situation best in getting a properly balanced account of events. Any difficulties in getting such a balanced account have been related to the task of selecting material specifically to illustrate the theme of the book, each of the parts of this book representing a condensation of reports of continuous work for periods ranging from fifteen months to two years. To give a picture of the various ways in which underlying forces affect behaviour in these industrial groups facing executive and consultative tasks, and to give a close-up view of tensions and difficulties lying behind selected day-to-day events which in themselves will be familiar to those in industry, has led inevitably to much greater emphasis on difficulties than on achievements. The normal, everyday affairs of the factory—the actual making of bearings, and the multitude of associated activities, such as selling goods, giving orders, estimating costs, undertaking research and development, and getting out accounts—have been omitted in the interests of following the main theme. The justification for this rests in the clearer presentation thereby obtained of the particular way of looking at the complex of factors influencing human relations in industry, with which the Research Team was concerned.

BEGINNING THE FIELD STUDY

Following the elaboration of these principles, the full Research Team was introduced into the factory in September. The team, as constituted at that time, was composed of eight members. These included the author and one other member of the Institute staff who was a psychologist with a considerable experience in industry and also in officer selection during the war. Along with these two, went six industrial Fellows, who had been specially selected for training and work on this project. Two of the Fellows were women—both graduates in economics, who had worked in the civil service. Of the four men, one had been a coal miner for eighteen years, one a fitter in an engineering works, one a draughtsman, and one a skilled engineering worker; three of the four had been trade union scholars at a University, and the fourth subsequently received a University scholarship at the end of the first year of the project; all four had held official trade union positions, and one had become a member of the National Executive Committee of the Amalgamated Engineering Union. Because of the working backgrounds of its members the Research Team mirrored in many respects the social composition of the factory. There are indications that this was of some importance in the acceptance of the team, although not of comparable importance with the demonstration in behaviour of independence of role and attitude, and skill in helping to work through problems.

The process of introducing the research workers to the factory was spread over several weeks, and included attendance at the normal two-day induction course for all newcomers, as well as attachment for short periods to the various departments. Following this introductory process, as it was highly unlikely that many sections would request help from the Research Team until its members were better known and their integrity tested-out, the months of October and November were occupied in carrying out a background study of the factory. This study was planned with the Project Subcommittee, and met the desire of the firm to have a picture of itself as seen by the Research Team, as well as giving an important opportunity to initiate the research by an overall appraisal of the organization.

The background study included a reconstruction of the factory's history and an examination of its social organization. History was looked at not as an academic study of the past, but as a technique for understanding how the factory had grown to be what it was, with particular emphasis on how events which had occurred in the past, and which had not been resolved, continued to affect the outlook both of those actually involved in the occurrence and of those who had heard about it at the time or afterwards. The examination of social organization was built around the assessment of how far the social structure of the factory had proved effective in coping with the forces which affected production and group relations. The results of this background study (given in Chapters Two and Three) were reported at a special meeting of the Works Council on 7th December, and, after discussing the report, the Council found it of sufficient value to endorse its previous decision to collaborate in the project, even though it was recognized that this might mean collaboration over a much longer period than had been foreseen.

The Projects Undertaken

The presentation of the background study to the Works Council concluded the first major phase of initiating the project and getting it under way. From that point on, there slowly grew sufficient confidence in the independence of the Research Team for a number of sections of the factory to ask for co-operation on specific problems; this is not to say that suspicion of the Research Team or apathy about the research disappeared for, as will be seen, such attitudes have affected the project throughout. The first two requests for Research Team assistance which were accepted actually came in October, two months before the report to the Works Council. The remainder of the requests were taken up between January and September, 1949. The first requests were of such a nature that by mid-January the whole of the Research Team was fully engaged with work arising out of such specific requests, and this situation has since continued.

A summary of the work undertaken is given below. None of these studies is in any sense completed, and work is continuing on each one at the time of writing, as shown in Diagram One.

Divisional Managers Meeting: A request from top management to help in sorting out relationships within the group has led to two years of collaboration with them at their regular weekly meeting in which aspects of executive leadership, policy making, and communications are being studied, and certain tentative

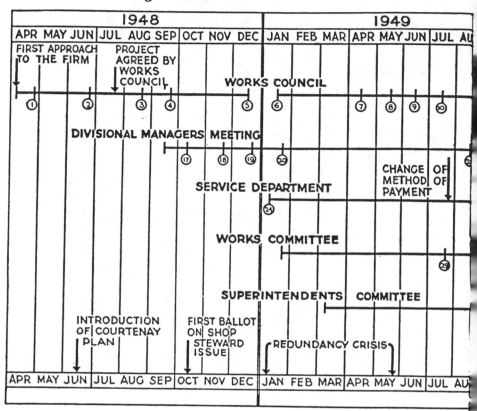

KEY TO CHART

INTRODUCTORY PHASE
   1. Management agree project
   2. Trade Unions agree project
   3. Introduction of Research Team
   4. Agreement of conditions of work
   5. First Research Team progress report

WORKS COUNCIL
   6. Beginning of sub-project on joint consultation
   7, 8, 9. Reports on Standing Committees
   10. Second Research Team progress report
   11. Week-end conference
   12. Beginning of re-consideration of Works Council structure
   13, 14. Meetings with Trade Union Officers
   15. First meeting of new Council
   16. Beginning of follow-up study

Additional studies of the Line Shop and the Finance Division, of labour turnover and of co mmunications are not shown on this diagram because they are not described in the text.

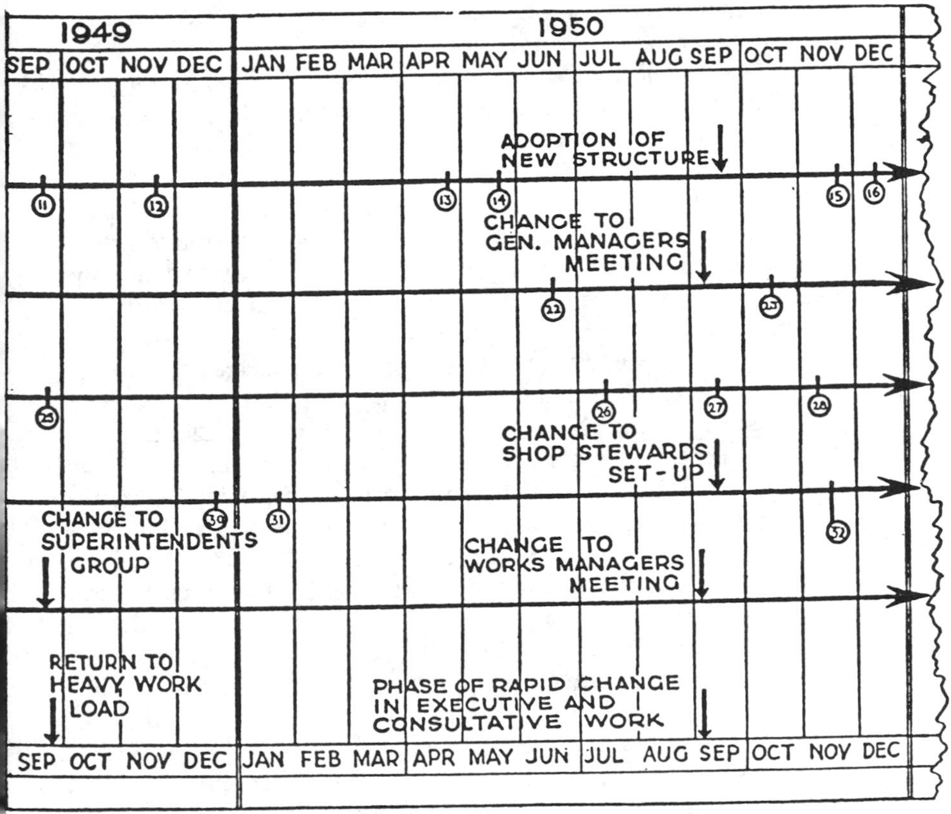

Diagram 1. The course of the project.

DIVISIONAL MANAGERS' MEETING
17, 18, 19. Special meetings
20. Beginning of attendance at regular weekly meetings
21. Clarification of the role of General Manager
22. Company policy document
23. New series of special evening meetings

SERVICE DEPARTMENT
24. Shop survey
25. First meeting of the Shop Council
26. Report on communications in the department
27. Changeover confirmed
28. Pilot study of principles and methods of timing jobs, with follow-up of effect of changeover

WORKS COMMITTEE
29. Discussion of personal cost of joint consultation
30. Report on communications
31. Start of full collaboration
32. First meeting of new Shop Stewards' Committee

conclusions reached about the effectiveness of executive management under varying conditions.

The Line Shop: A request from the workers' Shop Committee in this mass-production assembly-line shop, to help with some of its own problems has gradually led over a period of two years to work with the total department, including management and supervision, on group-bonus versus flat-rate methods of payment, relations between supervision and elected workers' representatives, and problems of relations between functional and line management in rate fixing. (This study is not dealt with in this book.)

The Service Department: A request from management, supervision, and workers to help in working through morale problems related to a change from piece-rate to hourly wage rate methods of payment. This has resulted in collaboration for nearly two years both with executives and elected representatives in the study of their relationships and relative power and authority.

The Works Council: A request from the Works Council for help with certain procedural difficulties has led to twenty months of collaboration in sorting out a wide range of problems including worker-management relations; the power, authority, and functions of the consultative bodies; and the difficulties in communication between the Council and the rest of the factory. This work has finally resulted in the emergence of an entirely new type of Works Council structure, and of a new conception of the Council's place in the policy-making activities of the company.

Superintendents Committee: A request from the middle management group to consider with them the problem of chairmanship of their group, led to eighteen months exploration with them of such problems as status and prestige and leadership at mid-management level. A change in structure took place, and a follow-up study of the new set-up is being carried out.

The Works Committee and Trade Unions: An invitation from the Works Committee to attend its meetings has led on to requests from this body and from a joint group, composed of management, shop stewards, and district trade union organizers, to help in working-through certain difficulties between the management and the unions, and between the workers themselves. This had led to over two years of co-operation in studying such problems as the workers' role in policy making, the structure of elected worker groups, and difficulties in communication between workers' representatives and their constituents.

Communications: Beginning at the request of the Works Council and continuing at the request of each level in the executive chain, an extensive study of difficulties in the operation of the channels of communication down the executive line has been made. To date, four of the factory's seven executive divisions have been covered.

THE COURSE OF THE PROJECT

The course of the project is shown in Diagram One. On this diagram the following points may be noted. More than three months, from April until July 1948, were taken up in getting agreement to undertake the project. The first sub-project was undertaken in September 1948 (the Divisional Managers Meeting), and all have continued.

As regards changes that have occurred, a definite pattern may be observed. The first main change took place when the Service Department altered its methods of wage payment in July 1949— just one year after the start of the project. This department also established a Shop Council which laid the basis for the pattern of organization subsequently adopted for the Works Council. The next change was in the composition of the Superintendents Committee, in September 1949. This change did not have a very deep effect at the time, but was a forerunner of the extensive changes which were to occur in the executive system.

It was only after more than two years of continuous work with the groups described that large-scale changes occurred— large-scale in the sense of affecting the factory as a whole. In September 1950 the Works Council adopted a new structure which extended the scope of democratic representation; the Works Committee became a shop stewards committee; and the top management and middle management groups adopted a straightforward executive form. These particular changes were the outward manifestations of profound changes in the firm's conception of executive and consultative processes. Since September 1950 the changes required in other parts of the factory by those which occurred in these central bodies, are in process of being worked through. As for the research project, follow-up studies have been initiated in connection with the consolidation of the changes.

CHAPTER TWO

HISTORY OF THE FIRM

1898 *to* 1949

THE Glacier Metal Company is a joint stock company with an authorized share capital of £500,000, and an issued share capital of £389,500. It employs some 1,800 people, of whom 1,300 work in the parent factory in London where this study is being carried out. There is also a factory in Scotland, with some 400 personnel, and a small Service Station in Manchester and in Glasgow. The company was founded in 1899 by two Americans who were interested in the manufacture of anti-friction metals. One of the partners soon after returned to America. Under the direction of the other, the firm slowly built up its trade and extended the range of its products, and in 1908 a move was made to larger premises. During the First World War Glacier manufactured the die-cast parts for hand grenades, and in response to the demands of the expanding machine industry, which was mass-producing motor vehicles, aircraft, ships, tanks, etc., the firm commenced the manufacture of bearings. This was the beginning of the present manufacturing activities. It was, however, the demands of industry rather than the particular desires of the owner which were determining the structure of the firm. He remained primarily interested in metal. The original metal-making operation still continues in the White Metal Shop, a small section of the Die Casting Department, employing six people and turning out the original anti-friction white metal which is exported all over the world. Around it has grown up the larger manufacturing concern.

FROM METALLURGY TO ENGINEERING, 1914–1935

The impetus given by the First World War to the development

and use of engines—in motor vehicles, aircraft, ships, and tanks—
increased the demands for the firm's products. In response to
this demand an engineer was brought in as Works Manager, so
that for the first time the organization had in a key position a
man whose interests were not confined to the production of
metal. The tremendous technical development, characteristic
of the war years, laid the basis for the expansion of the newly
developed motor-car industry. One large firm approached
Glacier with a request for bearings for its new products. This
new development called once more for larger premises, and in
1923 the factory was moved to its present location at Alperton.
Further technical developments brought the introduction of
bronze-backed bearings. The firm was now on the road to
becoming a large-scale manufacturing concern.

In 1929 the manufacture of steel-backed bearings was introduced for which power presses were brought in. A greater
number of semi-skilled at the expense of skilled workers had to
be employed. This change coincided with the economic depression of the early 'thirties, and while Glacier fared better
than many firms, there was a considerable recession. By 1933
the labour force had dropped to about 250, and the factory
worked short-time for a period of six months. To cope with the
decrease in business, a drive was made to lower costs and prices.
The line shop principle of continuous processing was introduced;
jobs were broken down into simpler components; and every
effort was made to reduce labour costs by the introduction of
unskilled labour. This policy was made possible by the national
and local unemployment situation in the country. Families from
the distressed areas were drifting into London in search of work
and were glad to accept any wage, however low. It was not
unusual for a married man to be taken on at a rate of 10d. per
per hour, while 11d. was the normal adult rate.

It was under such conditions that the Amalgamated Engineering Union began to organize the factory. On 21st February,
1935, a mass meeting was called in a nearby hall to protest against
the unsatisfactory conditions. The general aims of the organizers
of the meeting were stated in a short leaflet which read:

Fellow Workers
*The management are reorganizing the plant, introducing the line
system. What is the idea? Is it for the workers' benefit?*

Over one hundred workers have been dismissed in the last few months. Young Girls and Boys have taken their place at scandalously low rates of pay.

At the moment, the management are prepared to give a shilling here, a shilling extra bonus there to keep you quiet.

HOW LONG WILL THIS LAST?

It will last until the management have worked out the line system.

> *Then—away go the rates on all jobs.*
> *Then—mass dismissal will follow.*
> *Then—there will be all Boy & Girl labour.*
> *Fight back while there is time*

COME TO A MASS MEETING & HEAR HOW YOU CAN HELP YOURSELF.

The Chequers Hotel, Alperton
at 6 o'clock Thursday, 21st Feb., 1935.

The ferment continued, and further meetings were held during the following months, when suddenly on 14th April a strike was declared, without any official formulation of demands to management, or official sanction from the union. This state of disorganization may be explained by the fact that a large part of the union membership was new and lacked experienced leadership. The owner of the firm, whilst not a hard-hearted man, refused to meet the strikers. He is generally described as being highly respected by management and workers alike, but rather aloof and reserved, and unwilling to become involved in discussion of labour relations. The workers stayed out for a week hoping to receive the official backing of the Amalgamated Engineering Union, but this hoped-for support never came, because the strike had not been officially called; and so the strikers, led by a core of older workers, started to drift back. At the end of ten days all were back at their jobs; all, that is, with the exception of certain "agitators" whom the management had dismissed, alleging that they had damaged property or had been involved with the police. The management appeared to have won an unconditional victory, but the strike came as a considerable shock to everyone in the factory. It was not possible to pretend that everything was well, and the hands of those who wanted to improve matters were strengthened.

DEVELOPMENTS IN MANAGEMENT METHODS: 1935–1941

The period between 1921 and 1935 had been one of intensive technological development and growth and in 1935 Glacier was turned into a public company, with the owner as its first managing director. Under him were the Works Manager, and a relative newcomer to the firm who gradually took charge of administration. These two men did not see eye to eye, and a split

POPULATION – LONDON FACTORY

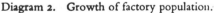

Diagram 2. Growth of factory population.

developed between the engineering side and the administrative side. This split was emphasized by the introduction of an accountant as company secretary who, in line with legal requirements to which the firm was now subject, made changes in the methods of financial controls, including more effective controls on supplies. These changes lent weight to the feeling which was growing that the "functional managers" were seeking to control the "line managers". In 1938 the Managing Director died, and his two subordinates were appointed co-managing directors, thus bringing their differences in technical and social outlook into the open. The difficulty was partly solved by the resignation of the older man, the engineer, in 1939, just one month before the outbreak of war. The split in the firm was not healed by this resignation, but continued in a more concealed fashion, the functional managers having come into the ascendant along with the remaining managing director.

The period following the outbreak of the war saw drastic changes in the size, composition, and location of the concern. A small No. 2 Factory was opened nearby in December, 1939, and shortly after, a carpet factory at Ayr in Scotland was taken over, and became No. 3 Factory. Some 80 members of the supervisory staff of all grades were sent to Scotland to develop the new works, which eventually employed 800 local workers. No. 4 Factory, a foundry, was established nearby at Kilmarnock to serve the new Scottish factory. In addition, two small shops, No. 5 and No. 6, were founded in North London. At the end of the war, factories No. 2, No. 4, and No. 6 were closed, while the relatively large No. 3 Factory in Scotland and the small No. 5 Factory in London were kept in production along with the main London works.

Between the years 1941 and 1944, particularly severe strains were put on the management structure owing to dilution, through expansion from about 1,000 up to 3,000 employees, and through the loss of expert management personnel to the forces. Under pressure of war conditions, these stresses were tackled on an emergency basis by increasing the controlling authority of the functional managers, such as the inspectors, the production engineers, the accountants, and the production controllers, who held responsibility for watching and checking on the line executives; a mechanism which intensified the feelings of a split in the management. This effect was somewhat mitigated by the recognition of the lack of a formally organized management structure. There was gradually evolved an organizational structure (produced as an organizational chart in 1944) based on the principle that no one should be responsible for more than some six to eight others, and that each person should know exactly to whom he was responsible. The possession of such an organizational chart was eventually to prove of inestimable value in healing the management split.

In 1941, in order to cope with the complex Government regulations covering employees and employment, personnel management was introduced as a management function, and a personnel manager was appointed. Some felt it was obviously the right thing to do, others, that it was but another instance of increased domination by specialists. Training activities were instituted, personnel record-cards were compiled, and all engagement of employees was transferred to the central Personnel Office, so that a more consistent personnel policy could be implemented. Towards the end of 1941 a well-equipped Medical Department was installed, with a half-time medical officer in charge, and a staff of two nurses. This came about originally as a move to cope with the dangers of lead poisoning, but quickly grew into a more comprehensive service.

THE INTRODUCTION AND GROWTH OF JOINT CONSULTATION:
1941–1947

In the same year, 1941, the management proposed that it should meet with representatives of the workers in order to discuss matters affecting the firm. This move stemmed largely from the initiative and interest of the new Managing Director, who was most keen on providing the widest possible opportunities for people to participate in the affairs of the factory. He therefore circulated, in December, 1941, a document which outlined proposals for initiating joint consultation procedures. Because it provides a direct impression of events at the time, a portion of the Managing Director's memorandum is here reproduced:

We (the Management) have come to the conclusion that until everybody in the factory has some voice in management we cannot be said to be doing our utmost to produce for Victory. We therefore propose that the non-staff and staff employees (excluding Grades I and II) should democratically elect from amongst their numbers a Works Committee composed of 24, of whom not less than 5 shall be women ... and that a Works Council shall be formed as follows: the Works Committee shall elect from their members four men and two women to form part of the Works Council; and five members of the Management as follows: one Director, the Works Manager, the Assistant Works Manager, the Production Manager, one departmental Manager representing the foremen in the Company. ... We want to make it very clear that we in no way disapprove of trade union members being elected to either of the two bodies it is

29

proposed to set up but that nothing done by the Works Committee shall interfere in any way with the right of the various members of the company to choose representatives as they wish by the exercise of their vote, and that under no circumstances shall membership of a trade union be a prerequisite of the right either to vote in the shop elections or to be elected to the Works Committee. . . . We most sincerely hope that because the Management are the instigators of the scheme, and because they propose to organize the first election, nobody will boycott the whole idea. We have a feeling that some people may react by saying "Put up show", "What is the motive behind all this?" or words to that effect. Because of this feeling we take the opportunity of saying now that there is no hidden motive in this move. . . . We believe that by the furtherance of the spirit of democracy within the Company much can be done to avoid injustices which by their effect on morale, reduce output and above all we hope that the inauguration of this scheme will make it easier for us all to become imbued with a proper enthusiasm—urge—spirit call it what you like, to work without reserve for Victory not only in the War but also in the Peace to come. Assuming, therefore, that you do in fact approve of this scheme, will you all please VOTE. . . . *Your shop representative will in fact be taking some part in the management of the Company, and to the extent to which you use your votes* YOU ALSO *will be taking part in management. Works Councils, committees, and trade union movements have in the past so often arisen out of the hard feelings created by inefficient or selfish management that many people still regard them merely as a weapon with which to* FIGHT *the management. They have frequently looked upon management as something which needs fighting all the time. We do hope that this Management's two years of office has sufficed to show that they do not require to be fought into giving a full measure of just conditions. What we want is your co-operation; we promise you that you shall have ours.*

This memorandum was widely discussed in the factory. Certain of the older employees were extremely suspicious, and some of the trade unionists saw the proposals as an effort to forestall the development of more orthodox trade union machinery. The foremen expressed their strong disapproval of the new plan in a discussion with the Managing Director, extending over two long meetings, for, was the dominance of the functional managers now to be supplemented by domination from below. In spite of these objections the Managing Director decided to go ahead, and this is one of the points perhaps where enthusiasm led to precipitate action, when a working-through of the differences might have been desirable. Working-through, however, would have been a formidable task, at a time when the concern was faced with bombing, the call-up of employees, and the opening of new factories.

The Works Committee came into being in December, 1941, and the Works Council in January, 1942. Six months later, the management suggested, and drew up constitutions for, setting up shop committees in each department. The intermediate grades of management, however, were by-passed, and information which they thought should have reached them from above began to reach them from below. There was anxiety that their authority was being challenged because of decisions that were, or might be, made by the Works Council. In short, management-worker consultation was established in the factory, without sufficient two-way consultation in the executive chain to allay the anxiety of the middle management strata. It was not seen at the time that effective joint consultation is dependent on the existence of an equally effective management organization with a high level of policy agreement.

The Works Committee and the Trade Unions

No initial insistence had been made that members of the Works Committee should be members of recognized trade unions, but the trade unionists, who were slowly increasing in strength, began to insist on this point. A warning was received from the Amalgamated Engineering Union, one of the main unions concerned, that it would take disciplinary action against its members if they continued to serve on a Committee which contained non-unionists. After numerous discussions which involved full-time trade union officials, an agreement was reached that membership of one of the recognized trade unions should be a condition of election to the Works Committee, but that this stipulation was not to be applied to the shop committees, and that voting for the election of representatives was to remain open to all.

Chairmanship of the Works Council

During 1947, the chairmanship of the Council was reviewed.

The Managing Director, according to the Council constitution which required that he appoint the chairman, had nominated himself to this position, in view of the strong desire from both sides of the Council to have his leadership. He did not feel able to speak freely, however, because of his double role of chief executive on the one hand and chairman of the Council on the other. In spite of strong representations from both workers and management, he decided to appoint another chairman. For this purpose he chose the Medical Officer; himself reverting to a position as a management member of the Council. To overcome the fears of his management colleagues, he promised that he would continue to maintain just as much interest in the Works Council as before; this having in any case been his intention.

The Medical Officer was not enthusiastic about being appointed chairman, and only agreed to accept on being urged to do so by the Works Committee members of the Council. A consonance of view between the management and the workers on the desirability of the selection of the Medical Officer, partly on account of his chairmanship skills and partly on grounds that he was the most "neutral" person in the factory, indicated that the Managing Director—in spite of deliberate attempts to avoid taking sides—had caused some stress by taking a more active part in discussions than was appropriate for a chairman, and had not been regarded as behaving in a totally impartial manner.

This attempt to solve a problem by an administrative change, directed largely toward the symptom, provided a measure of relief. But it did not wholly do away with the feeling that the management in general—and the Managing Director in particular—could still talk the workers out of anything; and although of some help, it still left the new chairman with the difficulty of coping with the causes of the trouble. Following the appointment of the Medical Officer in May, 1947, the constitution of the Council was changed so that the chairman was henceforth elected and not appointed.

GROWTH OF PERSONNEL AND TECHNICAL POLICY

A general picture of the factory between 1944 and 1946 can be obtained through the eyes of an industrial psychologist, a member of the staff of the National Institute of Industrial Psychology, who conducted a series of attitude surveys at the request of the Managing Director, who had himself persuaded his management colleagues that such investigations would be useful. Some of the findings of these surveys, based on interviews with over 250 employees, throw an interesting light on the organization during these years.

The psychologist found a striking freedom among employees to say what they thought without restraint or apparent fear. There was widespread agreement that the factory was a good place in which to work, and that it was steadily getting better. But having said this, and because of the security and good morale, they were able to voice opinions on many things which they wanted improved. Although joint consultation was encouraged, decisions on jobs were made without sufficient consultation. Employees considered they were given insufficient information by top management about company policy, and this had led to a growing belief that in fact there was no coherent policy. The feeling was strong that the management was frequently taking up new ideas and dropping them, leaving the concern a "graveyard of unfinished jobs". Concern over promotion policy was reflected in references to the high rate of turnover in senior personnel, and the felt failure to take technical qualifications into account in appointing executives. Education, youth, and intelligence were seen as more important than technical qualifications and experience. The only way to get ahead was to be a good leader, irrespective of technical qualifications. This was in part a reflection of the general criticism within the middle management ranks, that the rank and file had better contact with top management than they had. The exposure of these feelings contributed eventually to the extension of the joint consultative set-up by the establishment of staff committees, to allow the higher and middle management strata the opportunity for joint consultation.

Some indication of the mixed feelings in the factory towards the "new methods" is given in the rumour that the psychologist was passing confidential reports to the Managing Director at secret monthly meetings. Although this was not the case, it nevertheless illustrates how, in their desire for rapid change, the managerial personnel and workers' representatives, who took the lead in working for these changes, had been unable to take others with them as quickly as they would have liked. The Research

Team on the present project has had to work through some of the consequences of these difficulties.

Changes in Technical Practice

The development of the technical policy of the firm showed in another way how attempts were made to interest people in the broader aspects of industrial problems as affecting the organization, and to give them responsibility for their own immediate work. One important innovation to this end was the introduction of budgetary control methods, with decentralization of budgetary responsibility. This made each superintendent responsible for his departmental budget, and in theory at least made it possible for each section supervisor, and through him the worker on the floor, to participate in the economics of his own sphere of control. In actual practice the extent of delegation was tied up with each executive's assessment of the capabilities of his subordinates. In some cases expectations were raised but no increased delegation of responsibility occurred.

Another change in practice was in regard to inspection, where, due to the large numbers of new employees taken on during the war, it had become the practice to employ inspectors to check almost all the work carried out in the various processes. Supervisors and operatives felt that quality of work was something for which the inspector carried the responsibility. In 1944, the number of inspectors was cut from 102 to 56 within a period of two weeks. This reduction accompanied a policy change, which made the section supervisors, at their own insistence, responsible for quality, with the inspector in an advisory rather than a policeman role, a change which caused a certain amount of anxiety until it was seen that the scrap rate continued steadily to decline.

The use of consultative methods was also extended to new methods of production when early in 1946 a concerted drive was initiated to increase productive capacity through new and speedier machines and improvements in methods and designs. Since the introduction of such a plan meant turning out bearings at a higher rate than before, with re-timing and re-rating of jobs, the Works Director held mass meetings in each department, with frequent follow-up discussions between the Production Engineering Department and the shop floor, to explain the new methods and plans. The object was to let everyone know in advance what was going to happen, and to give those at shop floor level the opportunity to comment on these plans during their formative stages.

The question of participation in setting the policy regarding technical developments at various levels in the organization remains a big issue. The firm is very much alive from the engineering point of view and is constantly pioneering new products and production methods; and it designs and produces much of its most important machinery. The research staff with its mixed programme of pure and applied studies is in daily co-operation with the production shops on concrete problems, so that production personnel at all levels up to and including the shop floor have direct personal contact with the research personnel. This personal contact has not by itself removed the felt stresses between the shops and the research department, and it remains to be seen how far some of the current developments in clarifying the responsibilities and authority invested in the executive and consultative systems will go towards easing some of these stresses.

THE BEGINNING OF THE INSTITUTE'S RELATION WITH GLACIER

The historical events so far described have been reconstructed from documents and from interviews with long-service executives and employees. The following description of certain events occurring between 1946 and early 1949, is based upon first hand observation. The Tavistock Institute's relation with the firm began when the Managing Director took part, in 1946, in a weekly industrial discussion group at the Institute. In March, 1947, a member of the Institute staff spoke at the first Glacier Works Conference, attended by all managers from the Managing Director to the junior chargehands. Immediately after this conference the Managing Director decided that he wanted the Institute to assist the firm to overcome difficulties of which he was aware, but could only vaguely define. His divisional managers felt that he was still courting the danger of trying to get too far too fast, since the results of the work of the first industrial psychologist had not yet been fully implemented. These differences of emphasis between the Managing Director and his colleagues were sufficiently resolved to make it possible in June, 1947, for agreement to be reached independently by the Divisional Managers Meeting, the Superintendents Committee, and the Works Committee to ask the Institute, on behalf of the total factory, for assistance in dealing with outstanding problems related to the growth of the new social policy. (The discussions which led to this decision threw considerable light on relations within the management group, but this will be treated separately, in Chapter Eight.) A short study of factory morale was carried out by means of group discussions with a representative cross-section of the factory personnel.

A Study of the Factory

The first impression of Institute staff members, like that of the psychologist two years before, was that a remarkably sound morale existed in the concern. There appeared to be an overall sense of security, and a marked feeling of freedom to bring up those things which caused trouble or bother. Accordingly, in spite of the many difficulties that were expressed by the employees, the factory was described as a fine place in which to work, in which everyone was "treated as a human being", in a manner unfamiliar to many who had worked elsewhere. The difficulties that arose were broadly of two kinds: first, those run-of-the-mill difficulties which one would expect to find in any factory—problems arising out of normal day-to-day adjustments centred around wages, certain special inter-group relationships, and certain specific complaints, like dissatisfaction with the canteen; and second, those that were related more directly to the attempts in this particular factory to introduce a relatively new system of human relationships. We shall here touch only upon the second.

The second set of difficulties arose mainly because there had been inconsistencies in the manner in which new ways of running the factory had been introduced. Thus, for example, the management was made to smart for some years after an incident in 1946 caused by strong pressure from governmental transport authorities for Glacier to hold its summer holiday outside the August Bank Holiday period to help avoid travel congestion. Setting the date of the annual holiday (at that time, two weeks with one week's pay) had, however, become established as a matter to be decided by ballot, and it had been accepted as a foregone conclusion that the Bank Holiday period would be chosen. The Management, caught between strong governmental pressure on the one hand and the wishes of the employees on the other, sought relief from this dilemma by offering two weeks' holiday with *two weeks'* pay if the employees would vote to take their holiday outside the Bank Holiday period. This offer changed the vote in the desired direction, but was felt to be economic bribery, and, in spite of financial benefit, was taken as an example of the management's failure to adhere to its professed democratic policy.

It was also stated in many groups that rejection of a proposed spread-over wage scheme was based rather on objections to the way in which it was presented than on any dissatisfaction with the scheme itself. It was said in fact that the actual scheme had not been given serious consideration, and the occasion seems to have been taken as an opportunity to express some mild discontent over what were felt as management inconsistencies. Similarly, some of the lack of interest in the joint consultation procedures could in part be explained as arising from what appears to be the difference between the new programme, as put down on paper or described in talks, and the real substance of the programme, which was not felt to involve very much change. Some of the frustration which this discrepancy aroused was thought to have been expressed in the attitude of passive non-co-operation in joint consultation and other related procedures. In consequence of such happenings as those just described, the management's asserted policy of wanting to increase the opportunity for participation was looked at with a great deal of suspicion. The first question employees would ask themselves when some new scheme was mooted was "What's being put over us?" This questioning attitude showed itself in the very careful scrutiny that was given to every new proposal, and the out-of-hand rejection of any schemes in which it was felt that those in authority had not been completely consistent in what they started out to do.

The Institute consultants reported that the new policy had had the opposite but related effects of increasing satisfaction with the firm, while at the same time stepping up the level of aspiration for consistent democratic leadership so that actual behaviour was more rigorously criticized. This report was given wide circulation in the factory via key management personnel, elected representatives, and all who had taken part in the discussions. It was recommended that, if Institute collaboration was desired in tackling some of the problems raised in the report, discussions should be held between representatives of the management, the supervisors, and the workers, in order to determine how this might best be done. These representative groups, however, decided that they should try to work through their problems without any outside help, though it was agreed that any sections of the firm wishing for Institute assistance could ask for it.

A PROBLEM IN THE FOUNDRY: 1947

The first request for Institute assistance which was received after the above survey was completed is mentioned here because it throws light on some problems related to group bonus methods of payment. There had for some time been difficulty in the Foundry which came to a head in November, 1947, when the operatives downed tools for a few hours. The Works Director, who felt that an extensive examination would be necessary to determine the nature of the difficulties, obtained the agreement of higher management, of the Foundry management and supervision, and of the Foundry Shop Committee that Institute assistance should be obtained to examine the causes of the trouble. Group discussions with all Foundry personnel revealed a variety of felt problems relating to bonus payment, working conditions, the quality of supervision, as well as a sense of isolation from the main works derived from the feeling that the heavy manual type of work they did was regarded by others as inferior. We shall limit ourselves here, however, to the findings related to the bonus system.

The bonus system had been in operation since 1944. It consisted of giving a percentage bonus to the shop as a whole based on how much an estimated average production was surpassed; the production itself being measured in terms of the sales value of the goods produced. Each operative then received a bonus made up of this percentage of his own basic rate. In every group discussion the view was expressed that the bonus lay at the root of the discontent in the shop, the main criticism being that no one knew exactly how it was calculated. It was stated that although the system had been explained to everyone in the department, it was too complex to be clearly understood by anyone; added to which was the irritation that individuals could

38

not predict their own bonus, since they had to take all of the shop production into account and not just the results of their own effort. Variations in weekly bonus anywhere from zero (although this would be made up to 33⅓ per cent.) to 90 per cent., led to a considerable sense of insecurity about weekly earnings. The nature of the bonus led to arguments between the skilled and unskilled workers in the department, the skilled workers holding that they were entitled to a larger percentage than the unskilled, while the unskilled workers maintained that the bonus should be calculated as an absolute amount rather than as a percentage, since the skilled workers at higher basic wages were in effect getting a higher bonus each week than the unskilled.

One suggestion that came from many groups was that a higher wage rate with no bonus system at all would be preferable to the group bonus method of payment. Discussions about payment went on intermittently from that time until, in June, 1949, a change was made from the group bonus to a flat-rate system. This change—precipitated by increasing difficulty in agreeing rates due to technological changes in the department—was agreed after discussions between the management and the operatives and their shop and trade union representatives. It is too early as yet to state what the results have been. The general impression, however, of both the supervisors and the shop committee is that the new method of payment is more satisfactory. It has made possible the introduction of radical technological changes which have doubled the tonnage of the shop, while at the same time the operatives feel they are better off with their secure and fixed wage and the elimination of the continual wrangling over the size and uncertainty of the bonus. This is not to state that in principle flat rates are necessarily better or worse than a group bonus method of payment. It indicates, however, that the best method of payment is that which the people concerned agree is best at any given time; the real difficulty being to achieve this agreement.

THREE RECENT EVENTS: 1948–1949

To complete this historical survey of the firm, three different events will be described, two of which were sharply contrasting. Of the two contrasting events, one occurred as a result of a period of prosperity which led in 1948 to the implementation, by democratic discussion, of the policy of devoting a proportion of surplus revenue to raising salaries and wages; the other concerns a period of redundancy and crisis in early 1949 which threw up the rather more painful task of using the firm's democratic mechanisms to agree a policy to deal with the grave and harassing problem of laying people off. The third event, which took place between these two, during the second half of 1948, has to do with developments in trade union organization.

A Period of Prosperity

Late in 1947, the Board of Directors, having taken the total financial position into account, had decided that an additional £30,000 per annum could be added to the firm's wages expenditure, thus making it possible to implement the fourth point of the "Economic and Social Aspects of the Principles of Organization", namely, to raise wages and salaries when this was warranted by the surplus revenue of the organization. But how was this increase in wages and salaries to be distributed? The problem was referred to the Works Council, which promptly appointed a widely representative committee to handle the matter. This committee consisted of fifteen people, some of whom were Council members and others not, and including a divisional manager, a director, a departmental manager, supervisors, skilled, semi-skilled, and non-skilled operatives, and clerical workers. This committee, which became known as the "Gold Brick Committee" (the additional £30,000 soon became known as the "Gold Brick") asked for all members of the firm to send them suggestions on methods of distribution. It took some two months to work through the schemes which were suggested, and by a process of distillation, what finally became known as "The Courtenay Plan" (named after the member of the Finance Division who produced it) was put forward.

The Courtenay Plan was adopted by the Works Council, but only after some months of detailed consideration during which many modifications were made, and was put into operation in June, 1948. Under this plan permanent increases in wages and salaries ranging from 2s. 6d. to 10s. weekly were granted to all personnel, more or less in proportion to their total income. But not all the money was spent in wage increases. The plan also set out a method for increasing the security of the employees by the introduction of an extended sickness benefit scheme. This scheme provided for 88 hours sickness absence per annum with nearly full pay for everyone. The decision to extend the scope of the sickness absence scheme required the more general question of absence to be considered, and this led on to a discussion of time-keeping itself. At the instigation of the management, the basis of their discussion was broadened, with the unexpected result that a plan was adopted to abolish formal time-keeping in the form of clocking on and

CERTIFIED SICKNESS

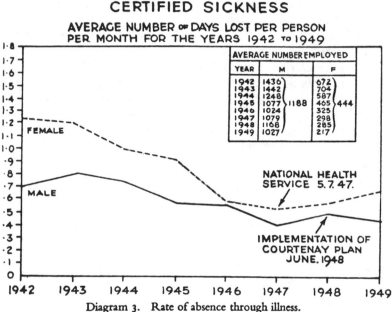

AVERAGE NUMBER of DAYS LOST PER PERSON
PER MONTH FOR THE YEARS 1942 to 1949

	AVERAGE NUMBER EMPLOYED	
YEAR	M	F
1942	1436	672
1943	1442	704
1944	1248	587
1945	1077) 1188	465) 444
1946	1024	325
1947	1079	298
1948	1168	285
1949	1027	217

FEMALE

MALE

NATIONAL HEALTH SERVICE 5.7.47.

IMPLEMENTATION OF COURTENAY PLAN JUNE. 1948

1942   1943   1944   1945   1946   1947   1948   1949

Diagram 3.   Rate of absence through illness.

clocking off for ordinary hours of work. Since then, employees have been paid their full morning or afternoon pay whatever time they come in, good time-keeping being a matter of personal responsibility, with section supervisors providing a final control in judging when such responsibility is not maintained.

As far as can be determined, the introduction of the sickness absence scheme and the ending of clocking on and off have had no marked effects in the direction of increasing either absence or lateness. It will be noted from Diagram Three that there has been only a slight increase in sickness absence since the sickness scheme was put into effect. This can be taken as one indicator confirming a generally high morale in the factory. Further confirmation of this is given in the lateness figures. The abolition of clocking on has produced no obvious increase in lateness, as measured by the gate-keeper, counting not individuals but total numbers of people coming in by stated times after the official beginning time.

Developments in Trade Union Organization

The second event to be dealt with in this section concerns certain developments in the organization of the Works Committee, The relations between this committee and the shop stewards' group in the factory had never been very clear, the Works Committee having included most of the leading shop stewards and having carried most of the functions of a shop stewards' committee, while the shop stewards as a special group had never seen fit to create an inter-union organization separately from the Works Committee. The matter remained relatively dormant between the time, in 1942, when the regulation was laid down that all Works Committee members had to be members of trade unions, and 1947, when the more active trade union members began to agitate for the Works Committee to be transformed into a shop stewards' committee, and for the factory to adopt a more conventional trade union set-up.

The ferment for change reached a high pitch in the summer and autumn of 1948. In July a series of mass meetings was held during works time at which the shop stewards and district trade union officials addressed the members of the firm and answered questions about the establishment of a shop stewards' set-up. This series of meetings was one of the outcomes of discussions between top management, the leading shop stewards, and the district trade union officials, in which the management offered the trade union officials all reasonable facilities for their attempts to achieve as high a level of trade union organization as possible.

The Research Team Project Officer was asked as an independent outsider to be chairman at these meetings, of which three were held in all, each attended by 200 to 400 people. On the platform were the Managing Director, along with the Works Committee Secretary, the London Regional Secretary of the Confederation of Shipbuilding and Engineering Unions and the district organizers of the five main unions, the A.E.U., T. & G.W.U., A.U.F.W., E.T.U., and A.E.S.D. The district organizer stated the case for the factory becoming a strictly trade union shop. Such meetings as these held in works time with the co-operation of management were, they said, undoubtedly an historic event, but the employees of the firm would indeed be adopting a selfish attitude if they said to themselves, "Everything is all right at Glacier, why should we bother joining the national trade union movement?" There was a great need for all workers to be members of their appropriate unions in order that a strong trade union movement could maintain sound working conditions throughout the country. The practical side of this could best be appreciated if they thought of it in terms of ensuring suitable conditions for themselves, if, as was likely for many of them, they transferred to other factories at some future time.

Following these meetings, a ballot on the structure of the workers' organizational set-up was arranged by the Works Committee for all hourly paid employees and members of Grade II and III Staff. The issue was as follows: a vote in favour of the existing Works Committee set-up meant that all members of the factory, whether members of a trade union or not, would continue to be able to take a direct voting part in electing their representatives on the Works Committee, and that anybody who was a trade union member could be elected as a representative; conversely, a vote in favour of a shop stewards' set-up meant that only trade union members could participate in the voting, and that qualification for membership of the Works Committee would be the holding of a shop steward's card, and not just the possession of trade union membership. Members of the Research Team who were at this time carrying out a background study of the factory, had the opportunity to discuss the matter with a large number of employees, and gained the distinct impression that there was a widespread lack of understanding of what the ballot was about, in particular of what would happen if there was a switch over to a shop stewards' set-up. Would a person also have to be a member of a trade union just to become a member of a shop committee, and would only trade union members in the shop be allowed to vote for their shop committee representative? Would the factory become a closed shop, or a union shop, or continue as it was, with trade union membership not necessarily being a requisite for employment in the company?

These and other questions remained unanswered, and, with this confusion still in the air, the ballot was held on the 21st October,

1948. A total of 60 per cent. of the votes had been set by the Works Committee as the necessary figure for any change to be implemented: the vote was exactly the reverse, with just 60 per cent. of the votes cast against change, and so the shop stewards' set-up did not come into being at that time. This ballot did not clear the matter, however, because those workers who were in favour of a change would not accept the vote as decisive. They argued that Grade II and III Staff members should not have been permitted to vote, since these groups, which were largely not union organized, were not directly affected by the results. They had their own elected committees and were not represented at that time by the Works Committee. Although the decision to include these groups in the ballot was made by the Works Committee itself, without any outside pressure, this did not carry any weight, mainly perhaps because those who did not accept the ballot were earnest trade unionists who aspired to see the factory completely organized, and who would carry on tenaciously until they had reached their goal.

During November, 1948, matters were taken a step further when the Grade III Staff Committee sought direct representation on the Works Council—a move which was opposed by the workers' representatives because it meant the possibility of having non-trade union members elected to the Council, and because they maintained that the Works Committee already represented Grades II and III, since personnel at these levels were allowed both to take part in the election of committee members and to stand for office on the Committee if they so wished. Indeed, they argued, the Chairman was himself a member of Grade III Staff. A special meeting was arranged between subcommittees of the Works Committee and the Grade III Staff Committee, and a compromise solution was thrashed out. This was that the Grade III Staff Committee should obtain its channel to the Works Council by electing three of its members—all of whom would have to be trade unionists—to sit on the Works Committee, which would enlarge itself from twenty-four to twenty-seven members in order to accommodate the change. The three new members were to be eligible for election among the nine Works Committee representatives for the Works Council; and in fact two of the three Grade III Staff Committee members were shortly afterwards thus elected.

There, for a time, the matter rested; the compromise having

44

dealt with the situation for the moment. It had, however, left unresolved many of the problems of the Grade III Staff members, as a group. One of the most important of these was the fact that Grade III included such a conglomeration of incompatible groupings, ranging from long-experienced foremen and section supervisors whose membership of Grade III represented a promotion from their former hourly paid status, to fifteen- and sixteen-year-old boys and girls who had Grade III status simply because they were office staff and hence on weekly pay. The three Grade III Staff Committee members on the Works Committee were all Section Supervisors from the works side of the factory, and the non-managerial office staff could be said hardly to be represented at all. Further, there was the paradoxical situation whereby junior managerial personnel sat on the workers' side of the two-sided Works Council. Because they were unresolved, the problems came up again when the structure of the Works Council came under review—an event to be treated in Chapter Five.

A Period of Redundancy

The Courtenay Plan was implemented in June, 1948. In January, 1949, the·economic situation radically altered, and the morale of the factory was sorely tested, when the world-wide change to a buyers' market adversely affected the engineering industry, and Glacier along with it. The intake of orders, and hence the amount of work on hand, diminished. The possible seriousness of the position was explicitly recognized by the management early in December, 1948, and was explained to the Works Council by the Managing Director at a special meeting early in January. The workers' representatives cross-examined the management. Why had not the situation been foreseen earlier and why had not measures been taken to prevent it having harsh effects on the factory? These big questions were only partially dealt with, and with everyone aware this was so, they turned to the urgent practical question of what they would do should the situation become so bad as to necessitate laying off employees.

In order to meet this situation if it arose, a committee composed of representatives of all grades of management, staff, and operatives was set up to study the problem. This committee produced a draft document which led to the adoption by the Council, later in January, of principles and procedures for coping with redundancy. The salient points were that if it was found necessary to reduce personnel through shortage of work, the personnel of the total factory and not only those immediately concerned, should be reviewed. Length of service, skills, and efficiency were all to be taken into account, and rough methods such as "last-in first-out" were not to be used. The final responsibility for deciding who were to be treated as redundant was to

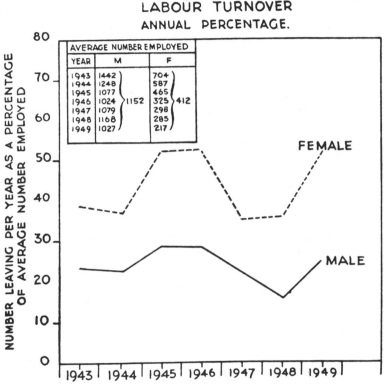

LABOUR TURNOVER
ANNUAL PERCENTAGE.

AVERAGE NUMBER EMPLOYED		
YEAR	M	F
1943	1442	704
1944	1248	587
1945	1077	465
1946	1024	325
1947	1079	298
1948	1168	285
1949	1027	217

Diagram 4. Labour turnover rate.

rest with the management. But employees' representatives were to be present at every stage while these decisions were being made, to ensure that the agreed principles were being carried out consistently in practice; and, as a final safeguard, every employee had the normal right of appeal. In addition to this, every effort was to be made during the period of notice to assist those leaving the firm to obtain employment elsewhere.

Then, in late January, the blow fell. It became necessary

for Glacier, along with many other light engineering establishments in the London region, to begin to lay people off. This continued until the employment position stabilized in April, by which time about 250 people, roughly 20 per cent. of the total personnel, had become redundant. In September, following the devaluation of the English pound, the engineering industry became very active, and the factory began once again to take on more people.

During the redundancy crisis, the operation of the agreed redundancy procedure was regarded on the whole as satisfactory. Interviews with those leaving showed that nearly all felt they

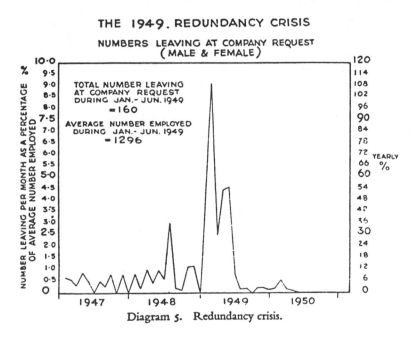

THE 1949. REDUNDANCY CRISIS

NUMBERS LEAVING AT COMPANY REQUEST
(MALE & FEMALE)

TOTAL NUMBER LEAVING
AT COMPANY REQUEST
DURING JAN.- JUN. 1949
=160

AVERAGE NUMBER EMPLOYED
DURING JAN.- JUN. 1949
=1296

Diagram 5. Redundancy crisis.

had been fairly dealt with. Whatever else their feelings, they were left with the impression of elaborate and scrupulous care having been taken in each individual case to ensure that no possible personal victimization could occur. The Works Committee sent a letter of thanks to one of the Personnel Officers who had been given the task of seeing all redundant personnel and trying to get them placed in other employment. As a result of his efforts, over 80 per cent. had got new jobs before they left the factory, and the Works Committee thanked him for the intensive effort he had put in to make this possible.

Satisfactory as the redundancy procedure was felt to be, there was, however, the inevitable result of great anxiety about job security. The procedure adopted had lessened some of the morale problems attendant upon redundancy, such as fears of victimization, but it did not, and could not, remove everyone's anxiety about the future and his own job. It was the first time under the new management policy that the factory had experienced such a setback in its fortunes and it had a deep and lasting effect. Those in leadership positions, both executive and representative, were left with the realization that they had created a factory way of life that was sufficiently tough and adaptive to survive adverse circumstances and to make a contribution towards coping with such circumstances in the most constructive way. The depth of the impact of having to lay people off, however, left no room whatever for any feelings of complacency. It took many months for the shock to wear off sufficiently for the problem to be talked about, and a policy hammered out that would go as far as possible towards minimizing the effects on the firm of the ups and downs in the economic forces acting on it. But this takes us well into the period after the research project had begun, and will be described in Chapter Five. It was necessary to be able to deal with redundancy problems if they arose; to prevent redundancy became a more pressing concern.

CHAPTER THREE

ORGANIZATION OF THE FIRM—1948

IN the following section the organization of the firm will be described as it was towards the end of 1948, when the background study was carried out by the Research Team. Minor changes which have taken place since then will be indicated. The larger changes which have occurred as a part of the work undertaken during the course of the project will be treated in subsequent sections, and the following description of organization will serve as a baseline against which to assess the extent and character of these changes.

THE LONDON FACTORY

The London factory is situated in the north-western industrial belt, a suburban area with a mixture of miscellaneous new industries and sprawling residential neighbourhoods; there is little homogeneity, and no organic relationship between the life of the factory and the life of the community round about. The employees live in widely scattered areas, so that even workmates in the same shop may be members of different branches of the same trade union catering for the craft. The split in modern society between home life and work is particularly marked in suburban industrial settings of this kind.

The firm manufactures plain bearings of all kinds—white metal, copper lined, bronze-backed—and of all sizes and shapes, from half an inch to two feet in diameter. The bearings are made by two processes. The first, which is used for the larger and more complex articles, begins with the casting of the bearing shells in the Foundry. These castings are then rough machined in either the Heavy Machining Department (B.1) or the Light Machining Department (B.2) and passed to the Die Casting Shop, where they are white metal lined, then back again to B.1 or B.2

for final machining. In the second process, used for the manufacture in large quantities of the smaller types of bearings (strip steel, anti-friction metal, and white metal-lined), steel strip is first of all made in the Strip-Line Shop, then passed to the Press Shop to be stamped into half-bearing shapes by high-speed presses of up to 250 tons. From here batches not exceeding 500 articles are routed to B.2 for machining; longer runs go to the Line Shop where the various machining operations, such as boring, milling, grooving, and flanging, are completed in a high-speed department with individual lines turning out up to 600 bearings per hour.

There is a Service Department to which worn bearings may be sent for relining. This department sells and installs replacement bearings as well as undertaking miscellaneous repairs. Under its control are the two outlying service stations at Manchester and Glasgow.

Serving the production and development work of the factory is a set of special departments, including a well-equipped Research Division which undertakes the developmental work necessary when bringing out new types of bearing, and is continuously engaged in improving metallurgical processes and engineering methods; a Tool Room, where most of the highly specialized tools are made; a Millwright Department, for maintenance and repairs; an Inspection Department, with Inspectors on attachment to each of the shops; a Production Engineering Department, which produces layouts and estimates; a Production Control Department, co-ordinating the flow of work; and a Drawing Office.

The work done is largely jobbing in character. At any one time there may be as many as 4,000 different kinds of bearing in production, ranging from special orders for one article to long runs of many thousands, so that planning a smooth workflow with even loading of sections and machines is an unusually complex task. The jobbing character of the firm's production has also created problems for the research project, in that the measuring of productivity and production efficiency under these conditions has presented such serious difficulties as to render them useless—up to the present, at least—as indicators of the effects of even gross social change.

FORMAL OUTLINE OF ORGANIZATION

The social structure and the code of behaviour evolved to cope with the demands of forces acting on the factory is stated in a document drawn up by the Works Council in 1946. This document, entitled *Principles of Organization*, is available in printed form for everyone. Worked out jointly by management and workers, it gives a clear indication of the constructive goals of those in leadership positions; but, as is so frequently the case with such documents, the extent and quality of implementation varies in different parts of the works. The principles are stated in five parts: principles governing the formal outline, the manning, and the economic and social aspects of organization; a code of industrial justice, and principles of consultation. These five parts are presented in italics in the following sections.

1.   *There shall be one Chief Executive in every organization who shall be responsible for carrying out the policy of its policy-making body. He shall be a full-time worker in the organization, and shall have complete authority to take any action consistent with the policy he is implementing.*
2.   *All reasonable steps shall be taken to explain to the members of the organization the responsibilities of its executives in general terms.*
3.   *The span of control of a "Line Executive" shall be limited to the number of people with whom he can maintain frequent contact, and amongst whom he can maintain co-operation. He shall grant the right of frequent access to those immediately responsible to him.*
4.   *The functional authority of all those carrying responsibility for special functions shall be clearly explained to everyone in the organization.*
5.   *Specialists or functional managers shall, within the span of their special knowledge, have the right to prescribe to executive managers on methods and techniques. They shall have the right to appeal to the executive manager, to whom they are themselves responsible, to endorse their prescriptions with executive authority, if those to whom their prescriptions are given fail to carry them out.*
6.   *No man shall be executively responsible to more than one person.*
7.   *No man shall give orders to anyone except those who are his immediate subordinates, emergency conditions excepted.*

8. *Diagrams clearly showing the relationship of people to authority and of one executive to another shall be exhibited in the organization.*

The Executive System

The present executive structure (shown in Diagram Six) was established in its main outlines by 1944. At the top of the executive hierarchy is the Board of Directors, composed of seven members, of whom all except two are full-time members of the company. One of the two part-time members, the Technical Director, had been a full-time member and maintains a continuing role as adviser on technical development. Both he and the other non-working Director take a personal interest in the firm's policy of social development. The other five board members are the Managing Director, who is Chairman of the Board, the Commercial Director, the Works Director, the Company Secretary, and the Personnel Director, all of whom are members of the London factory top management group. The Board retains full control over financial policy and capital expenditure above £25 and determines the main lines of development, but delegates responsibility for the day-to-day work of the company as a whole to the Managing Director.

The London factory is arranged in seven divisions—Service, Commercial, Finance, Personnel, Works, Technical, and Medical —each with a Divisional Manager at its head. The General Manager and his Divisional Managers together with the Works Manager meet weekly to organize and integrate the operation of the London factory. And at each level in the executive system, down to the section supervisor, it is intended that there should be meetings of each superior with his subordinates at least once monthly, although there are wide variations in practice in regard to this. This organizational structure, which has been systematically worked out and kept up to date, lays down the lines of authority and has given people a sense of having a definite place in the organization.

The differentiation between functional and executive authority had helped to clarify and define relations between executives. The present study raised doubts whether this differentiation was by itself sufficient if not accompanied by a precise definition of the relative authority of the various executives, since in practice the character of relations among executives is as much, if not more, determined by the quality of the relations among the people concerned as by the organizational structure and principles. As will be seen later, the distinction between "functional" and "executive" authority was eventually discovered to be inaccurate and misleading, and was completely discarded.

The Grading System

Within the executive structure, there operates a grading system. This system has grown up and developed over a period of twenty years, and does not represent any clear-cut differentiation of the executive chain. The grading system has become fairly complex with many features which are felt to be anomalous, because grade-status is given for reasons both of personal ability and of establishment.

There are forty members of Grade I Staff, made up mainly of divisional managers, departmental managers, and other senior personnel; and roughly eighty members of Grade II Staff, made up of departmental superintendents, senior secretarial staff, and leading technical and office personnel. There are approximately two hundred and eighty Grade III Staff personnel. This grade consists of particularly diverse membership, including office and clerical personnel, most of the section supervisors in the works, a small proportion of the skilled operatives, and technical personnel, such as draughtsmen. Certain benefits go with staff status. Grades I and II are paid on a monthly, and Grade III on a weekly basis. They have six weeks' sickness benefits, a higher geared pension scheme, free teas, a special annual bonus; while Grade I has an extra week, and Grades II and III an extra two days, of holiday.

MANNING THE ORGANIZATION

1. *Only decisions made by the Chief Executive shall be regarded as final within the organization. All other decisions shall be subject to the right of appeal up to the level of the Chief Executive if necessary. Final decisions made by the Chief Executive shall be subject to appeal only to agreed outside bodies, or to bodies which form a part of the Social Code.*
2. *Provided that it does not conflict with the policy of the organization the opinion of the executive immediately responsible shall be regarded as of major importance when considering the suitability of an individual to fill a particular post.*

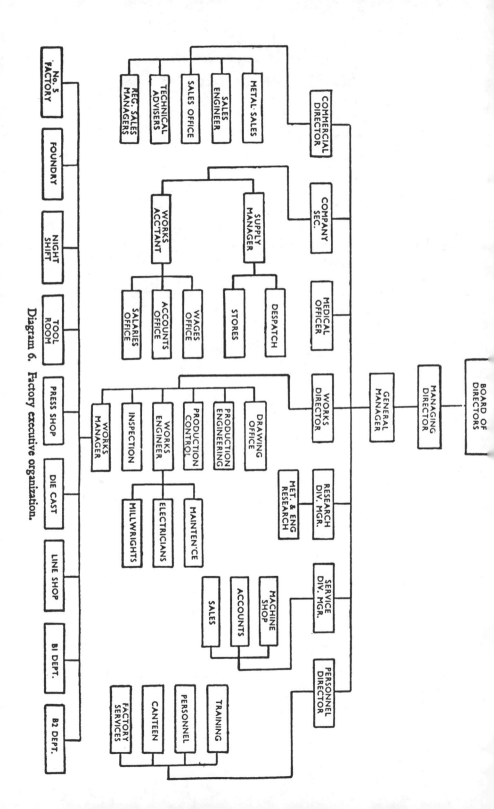

Diagram 6. Factory executive organization.

3. *Recognizing that the quality of executives selected has a substantial effect upon the happiness of many people, such selection shall employ the most scientific methods available. The probable reactions of his equals and of his subordinates, shall be taken into consideration by the executive responsible for selection.*

4. *Before resort is made to potential candidates outside the organization, vacancies which constitute promotion shall be filled from the body of existing employees, so long as suitable candidates are available within the organization. All such vacancies shall be advertised within the organization.*

5. *The excellence of a man's performance in his existing job, or the absence of a suitable replacement for him, shall not cause Management to refrain from offering promotion to posts for which he is suitable.*

6. *It is recognized that frustration is caused by the placing of men in jobs which leave their mental faculties under-employed. Selection procedures will endeavour to find people whose mental and physical calibre is neither too low nor too high.*

7. *Every executive shall draw up at least once every year, a list of recommendations on each of those immediately responsible to him, and shall discuss this report with his superiors. The main purpose of such a review of each individual, is to ensure that merit or insufficiency will not be overlooked.*

8. *An executive has, subject to the right of appeal, the authority to dismiss a man from his own team. Before taking this step it is his duty to confer with the Personnel Department, in order that they may explore the possibility of providing alternative occupation. The man will not be regarded as having been dismissed from the organization as a whole, until this attempt to find other jobs by the Personnel Department has failed.*

9. *Employees generally, through their elected representatives, shall have the right to make recommendations regarding the promotion of people within the organization.*

The Personnel Division is responsible for the engagement of all employees. Each new starter is interviewed, a personnel card (confidential to the Personnel Officers) is made out, and he then goes through an induction training course lasting a day and a half, during which the rules, regulations, and the ways of the firm are described.

Employee Report Forms have been introduced to assist in

keeping records of the abilities and qualities of all employees other than Grade I. These forms are filled in once a year by each employee's immediate superior, shown to and discussed with the employee, and then signed by him, but any employee who wishes has the option of not having a written report made out about him at all. The more complex problems thrown up by the introduction of such a progress report scheme were not foreseen, however, and there has developed an increasing resistance to the completion of the forms. This resistance, expressed mainly as criticism of the report form itself, seems to arise chiefly because all the difficulties in inter-personal relations between superiors and subordinates get thrown up once each year in connection with the filling in of the forms.

Advancement is based on a general policy, agreed by management and workers, of promotion from within. With few exceptions all vacancies are first advertised inside the firm, and only if no suitable candidates are found are positions advertised outside. The selection for positions of responsibility from section supervisor up is carried out by means of selection boards organized by the Personnel Division, usually composed of four persons: a representative of the Personnel Division, the executive responsible for the section in which the vacancy occurs, someone of the same rank and status as the vacant position, and an executive chosen from another department. The selection board procedure is based on an adaptation of the methods developed during the war for the selection of officers, and lasts anywhere from a half-day to two days. The procedure is made up of group discussions, individual interviews, and special selection tests related to the requirements of the post; and the names of those candidates who are short-listed are submitted for comments to the group whose leader is being selected. The Personnel Division functions in an advisory capacity only, the final responsibility for selection resting in the hands of the Board. The executive who will employ the candidate has the power to reject any candidate, but must have the agreement of the other members of the Board in making a selection.

Wage increases for hourly-paid workers and junior staff members are effected by departmental heads on recommendations from supervisors. Other increases are effected by the General Manager at an annual salary review carried out at the time the Progress Report Forms are completed. The Personnel Division reviews all wage increases and has the responsibility for making recommendations to the General Manager which will contribute to maintaining a balanced wage structure for the factory as a whole. Until 1940, rule of thumb and intuition based on experience had been the main methods used for maintaining a balanced wages structure. Since then, increasingly conscious planning has gradually been introduced into the wages structure, and some forward planning made possible, although anomalies still tend to occur and to be recognized only after there is trouble and complaint.

The Training and Upgrading Department of the Personnel Division has three full-time staff members. Its activities were inaugurated with lecture courses for senior management, and an extensive "Training Within Industry" programme has been carried on for the past four years for all executives, superintendents, and supervisors. The T.W.I. courses have been supplemented by a four-week, part-time course of individual training on the job, discussion groups, lectures, and visits. The Training Department is also responsible for the full training of apprentices and juveniles, and for a variety of special courses at night. These range from general education classes in English and mathematics to more specialized technical classes related to light engineering work, conducted by members of the firm.

The Medical Officer, in addition to his duties at the main works, supplies consultant medical services to a group of surrounding factories. He has under him a staff of three Nursing Sisters and a physio-therapist. The Medical Department is equipped with X-ray, and with minor surgical and physio-therapy treatment facilities.

ECONOMIC AND SOCIAL ASPECTS

1. *So long as shareholders are in receipt of a reasonable return on their investment, the surplus revenue of the organization shall be spent only in the five following ways:*

   (i) *Research and development*
   (ii) *Betterment of working conditions*
   (iii) *Betterment of equipment*
   (iv) *Raising of wages and salaries*
   (v) *Lowering of the price or raising the quality of the product.*

The organization shall strive unitedly to increase surplus revenue. The order of spending on different objectives must be settled by circumstances which will vary from time to time.

2. *The efficiency of management shall be judged not only by the profit made but by the following yardsticks also:*

(i) *Output per man-hour compared to the results achieved by organizations carrying out similar tasks.*

(ii) *By comparing the theoretical potential output with that actually achieved.*

(iii) *By comparison of the "labour turnover" of the organization with that of other organizations working under similar circumstances.*

2a. *The financial result of changes in method which increase efficiency or output shall redound to the credit of the organization as a whole and not in particular of those who devise or operate the improved method.*

3. *Financial and statistical statements comparing results achieved with those budgeted (standard costs) shall be available to all executives, whose sphere of responsibility is large enough to warrant its treatment as a cost centre.*

4. *Variations in rates of pay and salaries shall take into account the results of as scientific as possible an assessment of the following factors: (a) the value of the job being done by the individual to the organization; (b) the value to the organization of the personal qualities of the individual; (c) the past usefulness of the individual; (d) the potential usefulness of the individual. This principle involves the conception of equal pay for equal work, equal past usefulness, equal present reliability, and equal potential value.*

5. *Every employee shall have the right, assisted by his representative if he so wishes, to examine the calculations which form the basis of his output standards.*

Since 1941, the dividends paid to shareholders have been kept constant at 7½ per cent.; while this has not been declared as a fixed policy by the Board of Directors, the general expectation in the factory is that except in very special circumstances dividends will remain fixed at this level. It will be appreciated that this is a feature of considerable significance in all the developments in this firm towards industrial democracy. The surplus revenue shown in Diagram Seven has been spent since 1941 in the five ways listed, the Board of Directors taking the decision on how the surplus revenue is to be allocated. A description of the implementation of a company-wide wage increase is given in Chapter Two.

Although accurate comparative data are difficult to obtain, a rough survey showed that Glacier rates are about average for other comparable large factories in the district. There are a few with somewhat higher rates, and a few with lower, but the deviation on the whole is not great. All these firms are paying minimum rates considerably higher than the negotiated minimum

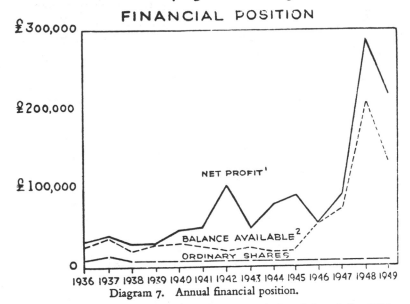

Diagram 7. Annual financial position.

(1) Net Profit — Profit remaining after making provision for depreciation and all expenses.
(2) Balance Avail. — (Net Profit—Balance brought forward from previous year) less (Amount payable under Employees' Profit Sharing scheme, Income Tax, Excess Profits Tax and Preferential Dividends.)

rates for the engineering industry; in the case of Glacier the excess ranges from two shillings to nearly two pounds per week.

The methods of payment are diverse. Staff members are paid on a weekly or monthly basis, and receive a special annual bonus, the size of which is determined by the Board of Directors in the light of the profit position and trading prospects. The total bonus is divided among the various staff grades in the ratio of four to three to two; Grade I Staff receiving double the bonus of Grade III Staff and 50 per cent. more than Grade II. All but a small number of the most skilled operatives are paid on an hourly basis.   Of the hourly-paid workers, about half, made up of the skilled workers and certain unskilled and semi-skilled workers, are paid a fixed hourly wage; about one-quarter are on individual piece-rates; and the other quarter are on small group bonus.   As will be seen in Table One, however, there is a trend towards flat-rates and away from piece-rates.   For the purpose of fixing rates, production engineers, part of whose job is to set times, are assigned to each production shop.   On all new jobs coming through, the rate fixer consults as much as possible with the operatives who will have to do the job. While this has had some effect in improving feelings about the fixing of rates, it has certainly not removed all of the complaints and difficulties connected with the size of the pay packet.

Method of Payment	*January, 1949*	*January, 1950*
Monthly salary	3	3
Weekly salary	26	26
Hourly flat rate . . .	36	46
Total: non-payment by results .	65%	75%
Individual piece-rate . . .	17	14
Group bonus—small group .	11	11
Group bonus—department .	7	0
Total: payment by results . .	35%	25%

Table 1.   Methods of Payment.

The percentage of personnel paid by various methods at the beginning of 1949 and 1950; the trend away from payment by results is shown.

There is a pension scheme for operatives with over three years service, two weeks holiday with pay, and a sickness benefit scheme which supplements the National Insurance benefits, and which allows up to eighty-eight hours of illness a year at slightly less than full pay.

The extensive social and welfare activities of the factory are organized and administered by the Social Club, which in turn is run by a committee whose Chairman, Secretary, and Treasurer are appointed by the company, the other members being elected each year at an Annual General Meeting open to all members.

The Club is financed through a weekly levy of 3*d.* on each employee, the revenue from its activities, and by special grants received from time to time from the company.

There are four canteens in the factory. Two of these are large canteens for workers, of which one supplies sandwiches and light meals, the other a full course dinner. There is a smaller canteen for Grade II and III Staff members and a still smaller one for Grade I Staff members. Separation of these canteens from each other had frequently been felt to establish a certain amount of status distinction, since staff membership confers the right to eat in one or the other of the two rather better appointed staff canteens at a slightly higher cost, although the same food is served in all.

The Social Club administers a sports field owned by the company and adjoining the factory. One of the canteens is converted into a gymnasium in the evening. Around these sporting facilities have been organized soccer, cricket, tennis and badminton teams and a boxing club. The boxing club and a boys' club which is also run by the firm have grown considerably since membership has been thrown open to families and friends of employees.

Supplementing the sickness benefit scheme the Social Club runs a Benevolent Society. This society has certain funds at its disposal which it is able to dispense at its own discretion to employees who run into special difficulties such as those arising out of extended illness.

CODE OF INDUSTRIAL JUSTICE

1. *Each member of the organization shall have the clear right of appeal to successive stages of higher executive authority and eventually to the Chief Executive. Every effort shall be made by all members of the organization: (a) to encourage individuals to use these channels of appeal; (b) to prevent victimization of those who use them; (c) to deal with appeals as speedily as possible.*

2. *(a) At any stage an appeal may be referred by either party to the Personnel Department for impartial recommendations to the appropriate executive authority and/or the appellant. (b) The Personnel Manager or appropriate Personnel Officer shall have the right and the duty to refer the appeal to the Chief Executive if his recommendations are not satisfactory to both parties.*

3. *There shall be an Appeal Tribunal, composed of one representative of management, one representative elected by the employees,*

and an independent chairman, to which purely personal issues may be taken if satisfaction through normal channels is not obtained. The majority vote of this Tribunal shall be final within the organization. This Tribunal shall be considered an outside body in reference to Clause 1 of "Principles of Manning".

4. When an appeal is made: (a) the appellant shall have the right to have present to help in the presentation of his case any person he may choose; (b) the person against whose judgment the complaint is lodged shall be present as well as the appellant. The decision of the executive to whom the appeal is made shall be given in the presence of the appellant.

5. When an executive finds it necessary to make such a report on one of his subordinates to his own superior as may prove detrimental to the future career of that subordinate, he shall inform the latter of the criticism made.

6. An executive shall not make personal criticism of a subordinate in the presence of others.

If an operative or staff member has any cause for dissatisfaction, he first sees his immediate supervisor. If the matter is not cleared up, he then sees his Shop or Staff Committee representative, and together they take the matter up again with the supervisor. If again the matter is not cleared up, two possible alternative courses are open. If it is a matter of wider interest, it may be taken to the Works Committee or Staff Committee, and from there, if it is an issue of sufficient importance, to the Works Council; or else the issue may be referred to the Departmental Chief. If he cannot resolve it, it may then be referred to the Personnel Division for advice, and a number of appeals are settled at this level merely by clarification of the regulations. If this does not happen, the matter may then be referred right up to the Managing Director for judgment. In normal circumstances the Managing Director is the last court of appeal. In arriving at his decision he takes into account general policy as agreed at the Works Council. If, however, dissatisfaction remains, there may be recourse to a special independent appeals tribunal composed of three members —one nominated by the Management, a second nominated by the Works Committee, and a third independent member nominated by a legal society. Although appeals as high as the Managing Director are not infrequent no one has as yet asked to make an appeal to this independent tribunal.

The appeals channels are used sufficiently freely to indicate that employees in general feel "safe" with the firm. Because of this security, appeals are often made to the Managing Director with the conscious purpose of testing principles and building case law; and as new principles emerge sufficiently clearly to be put into words, they are put before the Works Council for discussion and ratification. This is not to say, however, that fears of victimization have completely disappeared. Such fears persist side by side with the high degree of freedom in using the channels of appeal as shown at the Works Conference in 1949, when fear of victimization became one of the main topics of discussion; it was not described as a fear that someone higher up would purposely take it out of a subordinate, but rather that it was inevitable that superiors would unconsciously behave unfairly to those they did not like, even though they tried not to do so.

It may be said, therefore, that the appeals system has removed the fears of gross injustice which are so common in industry, but that the firm is now realizing that underneath these gross fears there are feelings which are more difficult to eradicate, and which were not visible, or not perceived, before the fears of a grosser kind were removed. These currently emerging difficulties seem to be related to deeply inbred feelings of suspicion and mistrust, caused by past family and working experience, and affected by any stresses in the factory situation.

PRINCIPLES OF CONSULTATION

1.   *Each primary working group shall elect one of their number to serve with a secondary group of representatives. If warranted by the size of the organization, each secondary group shall elect to a tertiary group of representatives and so on. In this manner, there shall be built up a hierarchy of employee representation consisting of a representative person or body of persons at every level of executive authority.*

2.   *Formal Meetings between management and those representative bodies shall take place. Due notice of these meetings shall be given to all members of the organization, or part thereof, concerned, and they shall have the right to attend as audience whenever reasonably possible.*

3.   *Management shall use these bodies as consultants in the making of their plans, so that employees become participants in the plan rather than mere executants of it.*

4. *Management shall take every possible step to keep these representative bodies informed of the current situation and future plans of the organization. No facts shall be denied to them, unless it is clear that the interests of the organization or members of it would be damaged by their publication. It shall be the duty of each representative body to keep all the workers whom they represent fully informed with regard to all transactions at meetings.*

5. *The representative bodies shall be encouraged by all possible means and safeguards to present freely their criticisms, suggestions, and aspirations, and those of the persons they represent.*

6. *Representative bodies whose advice is rejected shall have the right to have the same subject discussed as between the next higher level of executive authority and the corresponding representative body of employees.*

7. *An executive shall, when giving instructions or intimating decisions arrived at, convey to those he commands as much information as possible for the reasons for his decision.*

8. *It is recognized that, side by side with the formal means of consultation outlined above, there shall be at all times and at all levels of management and workers the fullest possible informal consultation.*

9. *Nothing in the foregoing 1 to 8 shall interfere with the right of the executive to make final decisions concerning all matters within the orbit of his responsibility, subject to the overriding authority of his own boss.*

The following description outlines the consultative structure of the firm as it was at the time the project began in September, 1948. Since then many changes have taken place, as will be outlined in subsequent sections of this report.

The consultative organization is pyramidal in outline. Every year each shop elects a committee to represent it, with one representative for each six to ten operatives. The function of the shop committee is to consult with the supervisory staff on the day-to-day running of the department. This shop committee structure is intended to give all operatives a feeling of direct communication with the consultative hierarchy, and to this end the committees meet at least once a month. There are a number of unresolved difficulties, however, which have not yet been tackled. One of these is that in most shops the shop committee members are not elected by sections, but by the whole shop, with the result that there may be two or more shop committee members in some sections, while others will be without any, so that some sections may be assigned a representative who does not work with them. Even with a different electoral system it would be practically impossible to arrange for shop committee members to represent the groups with which they work, because people are moved about so much from machine to machine, from section to section, and from department to department. As a result, it is still questionable whether the shop committees really give any more direct representation or channels of communication to operatives than merely having one or two representatives or shop stewards for a department as a whole. On the other hand, in those cases where the shop committees seem to function well, it appears easier for a committee of six to ten to take major decisions on behalf of a large shop, than it would be for one or two representatives to do so.

In addition to its shop committee each shop elects one representative for approximately every fifty members to the Works Committee, which thus represents all of the operatives, and contains in addition three representatives elected from the Grade III Staff Committee. To be elected to the Works Committee, membership of a trade union is required, but non-trade union members participate on an equal footing with trade union members in the elections. The Works Committee meets at least once a month. It has a full-time secretary, elected by the Committee and paid by the company, whose job it is to assist the Works Committee and shop committee members to carry out their work, and to help in sorting out everyday problems at shop floor level. Some of the difficulties in running the Works Committee and in sorting out the roles of workers' representatives will be dealt with more fully in Part Two.

The Works Committee in turn elects nine members to the central consultative body, the Works Council, which also has nine representatives of management. The constitution lays down that these management members shall be the Managing Director, the Medical Officer, the Personnel Director, the Commercial Director, the Works Director, the Works Manager, a representative of the Superintendents Committee, the Service Division Manager, and the Company Secretary. The Works Council meets at least once a month, and holds occasional extraordinary meetings when there are urgent matters to be decided. The constitution and the operation of the Council will be described in more detail in Chapter Five, but one matter, the voting procedure, shall be mentioned here.

Decisions on all matters of policy must be supported by a unanimous vote, and not by a mere majority. This means that there must be agreement among all members of Council on all decisions of major importance. The principle behind this is that disagreements must be ironed out in advance of taking decisions, and that it is better to have no decision at all rather than a partially supported decision, and then spend the next months or years in a continuous wrangle between the satisfied majority and the dissatisfied minority. The operation of this principle has not noticeably held up proceedings. It has meant that whenever there has been disagreement, more study of the problem has been necessary, so that, in the light of new facts, new solutions, agreeable to all, can be achieved.

This principle of unanimity voting is now widespread in the firm. It is a principle of some importance, for it supports a conception of a community in which disagreements will be handled not by majority domination, nor by compromise, but by a working-through to new solutions by such exploration of factors affecting a problem that sufficient clarity to allow resolution is achieved.

The Works Magazine

The Works Council controls the works magazine called *Our Opinion.* This magazine is run by an editorial board set up by, and responsible to, the Council. It appears monthly; part of its income being derived from sale at 3*d.* per copy, and part from a financial grant from the company. The general lines on which the magazine is run have frequently been debated at Works Council, and the result of these debates has been to lay down a broad policy of using the magazine as a forum for the expression of opinion about the running of the firm. Its contents, therefore, tended at one time to include rather serious and weighty articles and letters to the editor on controversial issues. This policy, however, led to gradually diminishing sales and to a dissatisfied editorial board, which found it difficult to get together sufficient material to make their efforts seem worthwhile. By means of the consultative machinery it has since been possible to revise the editorial policy and to bring it more into line with the general interests of people in the factory, and the magazine is now rather lighter in tone, although first priority for publication is still held for those contributions dealing with serious and controversial company affairs.

Staff Committees

In addition to this pyramidal consultative structure based on the works operatives, each of the three grades of staff has its own elected committee. Grade I Staff elect a committee of five, which meets infrequently, since the main affairs of the Grade I Staff are handled by the Divisional Managers Meeting. Grade II Staff elect a committee of eight, and the Grade III Staff a committee of fourteen. These both meet once a month. All three staff committees have a direct channel to the Divisional Managers Meeting through the Personnel Director. The Grade III Staff Committee also elects three members (who must be trade union members) to the Works Committee. Thus, whereas Grade I and Grade II Staff Committees are represented on the Works Council through the Divisional Managers, the Grade III Staff Committee is represented on the Works Council through both the Divisional Managers and the Works Committee.

To complete the consultative structure, one other committee must be mentioned; namely the Superintendents Committee, which elects a representative to the Works Council. This is a body of seventeen members composed of all the Works Division Superintendents together with certain personnel at superintendent level from certain other divisions. Further description of this committee will, however, be left for more detailed consideration in Chapter Seven.

A plan of the consultative organization is given in Diagram Eight. It will be noted that there is a considerable difference in the consultative structure for hourly-paid operatives and for staff. The former are directly represented in their own shops, by one representative more or less on the spot for roughly each ten operatives. The latter, however, have representatives who cover rather larger geographical regions, and are neither so accessible nor so well known to their constituents. The staff are not so directly represented on the Works Council as the hourly-paid operatives, and this has had at least two effects. First, there is a general feeling that management is more willing to consult with hourly-paid operatives than with staff, and

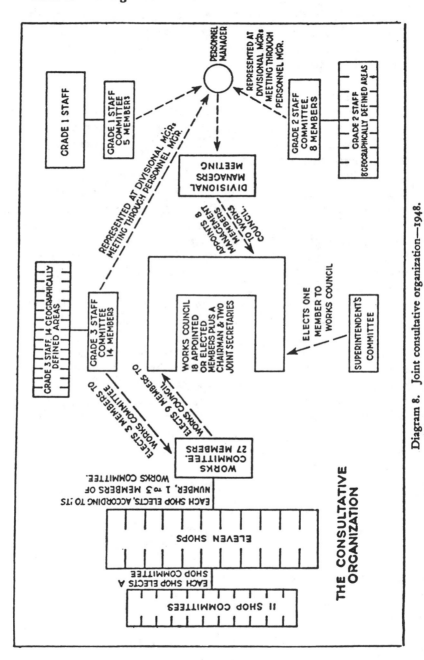

Diagram 8. Joint consultative organization—1948.

secondly, it has meant that the Divisional Managers Meeting, and not the Works Council, has been the central point at which inter-group relations in the factory have been sorted out.

This rather mixed picture of representation on the Works Council results from maintaining a two-sided worker-management structure with the consequent difficulty of deciding where to draw the line between worker and manager. At Glacier this line has, on different occasions, been drawn both above and below Grade III Staff, reflecting perhaps the general feeling that it runs somewhere through the middle of this grade. As will be seen in Part Two, however, the factory is moving towards a rather more complex form of representation on the Works Council.

Trade Unions

There are five main trade unions—the Transport and General Workers Union, the Amalgamated Engineering Union, the Electrical Trades Union, the Association of Engineering and Shipbuilding Draughtsmen, and the Amalgamated Union of Foundry Workers. There is a District Officer for each union, and a degree of co-ordination is achieved by common membership in the Confederation of Engineering and Shipbuilding Unions.

At the present time the hourly-paid operatives are somewhat over half organized and this has led to considerable complexity in the consultative structure. It has been mentioned that all members of the Works Committee must be members of a trade union, but this does not obtain for the shop committees. In the shops, therefore, the anomalous position arises whereby workers are represented both by their elected shop committee members and by shop stewards who may or may not be members of the shop committee.

In addition, although all of the Works Committee members are trade unionists they are not necessarily shop stewards, although a very high percentage do in fact possess shop stewards cards. As a result of this the shop stewards' group overlaps but is not identical with the Works Committee. This has led in turn to confusion between the responsibilities of the Works Committee members and of the shop stewards, both of whom are in many cases the same persons. Some of these difficulties will be treated in Chapter Five, in which trade union developments will be described.

PART TWO

THREE YEARS OF CHANGE

This part contains descriptions of work done in co-operation with five sections of the factory at their own request. Each chapter is a self-contained narrative, intended to provide the opportunity to share the events experienced by members of the factory as they tackled the problems facing them. No conclusions will be presented at this stage, the cumulative effect of the five separate studies being intended to give a general impression of what the life of the factory is like.

CHAPTER FOUR

THE SERVICE DEPARTMENT

Methods of Payment and Morale

WHETHER people work more efficiently and with greater satisfaction when there is a direct financial incentive, such as that provided by piece-rate systems, can be considered an open question. Indeed, it is unlikely that such a question can be answered by itself, because the way people are paid is only one facet of a large number of interdependent sociological, technological, psychological, economic, and cultural variables which interpenetrate to create social climate, and community morale in industry. An opportunity to study this problem in some detail occurred, when in January, 1949, the Research Team received a request jointly from the management and the workers in the Service Department to assist them with discussions on whether or not they should switch over from piece-rates to hourly wage rates.[1]

ORGANIZATION AND HISTORY OF THE DEPARTMENT

The Service Department is similar to a small company. It is a relatively independent unit engaged in the sale of replacement bearings and in repair work, with subsidiary manufacture of small runs or special orders. It has its own administrative staff, drawing office, and sales organization, employing altogether some 100 people, 40 of whom were at this time on piece-rates. Its organization and activities are shown in Diagram Nine.

It was first established as a separate department in 1931 as a result of an increased demand from customers for repair services. Until the war years the new shop felt itself to be separate from

[1] A preliminary account of this work in the Service Department appeared in *Human Relations*, Vol. III, No. 3, August, *1950*, pp. *223–249*.

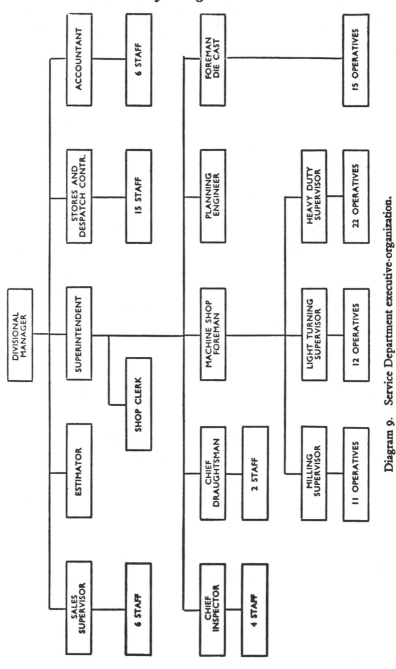

Diagram 9.   Service Department executive-organization.

74

the rest of the factory. It had its own customers, with whom there was close personal contact, for much of the pricing of jobs was done by direct meeting between customers, supervision, and operatives. This feeling of independence was fortified by the shop having its own gate and working different hours from other shops; its operatives were not asked to take part in the 1935 strike. During the depression years, its members felt more secure than other Glacier workers, because the shop was steadily growing and able to take on workers laid off in other parts of the factory.

During the war the work of the department had become so extensive that increasing systematization was introduced in costing, pricing, and handling of stocks, and this brought an end to informal contact with customers. Also during this period there was some lessening in the amount of repair work carried out, customers preferring to obtain new rather than relined bearings. This change necessitated modifications in the department's activities, and, in particular, with regard to their stores and commercial activities.

In 1943, a *Payment by Results on Time Basis Scheme* was introduced. This was a payment by results scheme with rates calculated in standard minutes, rather than a money contract for a given job which the operative could then complete as quickly as he wished. This new system was tried out for three months, at the end of which management was satisfied, but the workers were not so sure. Partly as a result of the attitude of their trade union officials, who pointed out that they seemed to be better off financially under the new system, the workers agreed to change over, on condition that they could change back if they wished. Although there is no indication that such a change was asked for, the *minute system*, as it became known, fell into disrepute, and the impression grew in the shop that it had been imposed by the management.

In 1947, the Shop Manager retired, and the present Divisional Manager and Shop Superintendent were brought in. They were most anxious, in line with the general policy of the firm, to establish good relations in the department, and to bring it into closer contact with the rest of the factory, but felt only partially successful. The workers' representatives, led by the convener of shop stewards of the Amalgamated Engineering Union, remained suspicious, not only of their own departmental management, but of the whole consultative set-up of the factory. They had withdrawn their Works Committee representatives in 1944 because they considered the Glacier model of joint consultation out of line with normal trade union practices, and had only consented as late as November, 1948, to elect representatives once again, for a trial period of one year, during which they intended to consider their position further.

THE NATURE OF THE PROBLEM

The proposal to change over to hourly rates was first mooted by the Divisional Manager in February, 1948, in a talk to the whole department, in which he reasoned that service work, which consisted of repair jobs, did not lend itself readily to payment by results, since no two jobs were alike and jobs differ each time they come through the department, because of distortions and varying conditions of the bearing shells. Piece-work prices as set on work of this sort could only be estimates, so that constant adjustments were necessary to ensure a fair rate. For a majority of jobs this meant a discussion to work out an adjustment on the existing rate, which not only used up time but involved complications in the costing and financial organization of the department. There had been continuous dissatisfaction with the piece-work system in use (the so-called *minute system*) ever since its introduction by the previous management in 1943; some jobs paid well, others not so well, with the result that it was possible for unskilled operatives to earn more than craftsmen. The chronic irritation produced by the system was believed both by management and workers to be costly in reduced output and in lowered morale.

The reaction of the operatives remained reasonably favourable to the proposed changeover at intermittent discussions held during the year, and the management agreed to get out proposals for an average hourly rate for piece-rate operatives, based on the average level of piece-rate earnings for the shop. On 31st December, 1948, the Divisional Manager called a meeting of all operatives and offered an average flat hourly wage of 2s. 8¾d.[1] to the pieceworkers. This figure included a deduction of about 1d. an hour per operative to allow for a possible slight decrease in productivity under a flat-rate scheme. Since this was an average some would earn less, and others more. The exact method for

[1] None of the payment figures quoted include national bonus.

determining each individual's rate had been left to be agreed between management and the workers' representatives.

During this meeting one question was raised which was to recur frequently, "How would output be maintained when piece-work incentives were withdrawn?" The Divisional Manager's opinion was that this was essentially up to the workers themselves, but he was confident that people would behave responsibly and that output would suffer little, if at all. Checks on the level of productivity would have to be made, he thought, but this could be done in broad terms and need not be related to individuals.

A *Wages Committee*, composed of the Shop Committee, the Divisional Manager, the Shop Superintendent and the Shop Accountant was set up to consider the matter in more detail. It held its first meeting on 2nd January, 1949, when the workers' representatives reported mixed feelings in the shop, with some in favour and some suspicious of the proposed changeover, the latter attitude expressing itself in such comments as "What were the management up to now?" and "What are they going to get out of this?" In view of the suspicions complete facilities were given to the Shop Committee Chairman to make whatever checks he liked. This included the provision of detailed department figures from which he could make independent calculations in consultation with his own trade union officials. It was also decided to enlist the co-operation of the Research Team both "to obtain advice on how to avoid likely pitfalls", and with the hope that the presence of outsiders might in some way alleviate some of their difficulties. In reply to this request, the Research Team Project Officer met the Divisional Manager on 5th January, and on behalf of the Research Team, agreed to provide consultant services.

NEGOTIATIONS ABOUT PAYMENT

The Shop Committee Chairman spent much of January and February reviewing the wages figures and consulting with his district officials. This review completed, the second formal meeting of the Wages Committee was called.

Wages Committee: 23rd February. As this was the first meeting the consultant had attended, he took the opportunity at the outset to explain his role as laid down in the terms of reference of the project. The management then explained that although the piece-workers, because of the penny an hour deductions, would lose about 2*s*.—4*s*. per week per person, this would be offset by their greater security. The workers would not accept this, because there were certain operatives who would lose materially. One person in the shop, who (it was admitted) was earning anomalously high rates, would suddenly be deprived of about £1 10*s*. a week; and another group of six or seven would lose between 6*s*. and 12*s*. In spite of these difficulties, however, the Shop Committee did consider that the proposed new set-up would be better for the shop as a whole; and therefore, provided that a satisfactory formula could be found to ensure that no one would suffer too much, it did not seem unfair that some should lose a little in order to achieve a better balanced wage structure for all.

During this meeting the consultant suggested that there might be value in finding out in more detail what the Shop felt about the new proposals, in order to take into account the operatives' feelings, not only about the wages question, but also about the morale issues which would inevitably be bound up with it. This suggestion was not discussed; instead a small sub-committee composed of the two Works Committee representatives, one Shop Committee member, the Divisional Manager, and the Departmental Superintendent—was set up to work out a fair method of calculating individual wages, so that no one would lose too heavily.

Wages Sub-Committee: 25th February. The Divisional Manager opened the meeting, saying that its purpose was, first, to consider how the Research Team could help to organize group discussions to enquire into attitudes towards the proposed change-over and, second, to discuss proposals for calculating individual wages under a flat-rate scheme. The Shop Committee Chairman immediately disagreed. They were there only to talk about wages proposals. The Shop Committee itself was the sole body which could take decisions on finding out what the workers on the shop floor were thinking. The consultant referred here to the Shop Committee's suspicion of himself and his role, and emphasized that he and other members of the Research Team would act only in concert with the Wages Committee as a whole. But this suspicion of him must surely in part indicate suspicion amongst themselves, and the fear that one party would be able to use him against the other.

This reference to existing attitudes in the group seemed to reassure them, and they went on to a discussion of various wages proposals. But, during the next hour and a half, not more than fifteen minutes were spent on the proposals themselves; the rest went on a wide variety of general morale issues such as : How much will the change-over alter the existing relationship among workers, between workers and management, and particularly between workers and supervisors? What happens if wages are fixed and production goes up? What techniques can be used to get general agreement in the shop? How will the supervisors behave under such a scheme? Can a supervisor be guaranteed that no worker on his section will earn more than he does?

When the consultant commented that their talk was demonstrating how inextricably the wages proposals were tied up with other morale issues in the Shop, the Divisional Manager suggested that each supervisor might be asked to enquire into the feelings of his section. The Shop Committee Chairman opposed this "investigation of workers' attitudes, since all supervisors were on management's side".

The Divisional Manager complained "You're suggesting there are two sides to the table. I feel that we're all in this together."

"There *are* two sides to the table, and I don't want the supervisors poking their noses into the Shop Committee's business."

"I don't think we can go on if you're going to use language like that."

"I don't care whether you object to my language or not. I'm going to be blunt; this is the way I feel about the matter."

The atmosphere was charged. There had recurred in a slightly different form what had earlier appeared as suspicion that the Research Team would usurp the Shop Committee function, and possibly act on behalf of management alone. The consultant therefore interpreted this to them as a displacement on to supervision of suspicion towards himself and the Research Team, representing once again their own suspicion of each other at the moment. This, however, was vigorously denied. But, quite unexpectedly, it was suddenly recognized that there was no representative of supervision on the Wages Committee, and they decided to remedy this shortcoming before the next meeting.

Shop Committee: 2nd March. A few days later, the consultant was invited by the Superintendent to attend a meeting in his office. Agreeing, the consultant found himself at a private meeting of the Shop Committee; the Shop Committee Chairman having arranged this with the Superintendent. The consultant having raised the issue of the suspicion in the Committee of the Research Team, the suspicion itself appeared to be diminished somewhat. But what, he asked them, was to be his role at such a meeting; did they wish his advice on how better to handle management; and what would they say if they saw him or any other member of the Research Team meeting privately with management, or with supervision? Surely, he said, if he would help them against management, they would have every reason to suspect that he might equally help management against them.

This last point struck home, and made it possible to settle two important questions before the consultant agreed to remain with them. First, he would comment only on their own relations with one another, and on their work as a Committee; that is, he would not put forward views on any persons or groups other than those immediately present. Secondly, they should raise at the beginning of the next meeting of the Wages Committee the question whether or not the Research Team might meet independently in this way with any section of the department—whether management, supervision, or workers.

During this meeting, the Shop Committee members expressed far more anxiety than they had dared reveal openly in the Wages Committee. The management, they feared, was trying to put something over on them, and if output went down, would scrap the whole scheme and put them back on piece-rates. They were uncertain what to do, because the people on the shop floor whom they represented were deeply suspicious of the proposals. Because, from their discussion, there was little indication that they had a rounded picture of the attitudes of their constituents, the consultant suggested they might find their job a lot easier if they would undertake serious discussions throughout the shop before proceeding further. This would not only acquaint them with shop attitudes, but would give the shop an opportunity for more active participation, and would make it easier for themselves to report back developments as they occurred. Although some continued to maintain that the shop was interested in nothing but the size of the pay packet, the Committee as a whole accepted the suggestion and asked for the assistance of the Research Team.

In a short special meeting the next day, the Wages Committee adopted the principle of the Research Team co-operating with sub-groups in the department, and the consultant found himself in a position he had not previously experienced—a relationship with a group composed of management, supervisors, and workers, and at the same time carrying on independent relations with the component parts. Role conflicts that might arise—as, for example, when meeting separately with the workers, he had already learned of attitudes towards management which had not been expressed in the Wages Committee meetings—he hoped to avoid by relying more than ever on the general method of confining his remarks to what was happening in the here-and-now of each group. Although it was impossible for him not to be affected in his observations of the total group by what he knew from contact with the parts, he anticipated that he could avoid confusion on this score by the ordinary procedure of always having tangible evidence in the here-and-now, under whatever condition, for the interpretation made.

ATTITUDES AT SHOP FLOOR LEVEL

Group Discussions: 8th March. The forty piece-workers were divided into five groups of eight, with one Shop Committee member delegated to each. To obtain a comprehensive picture it was decided to have one of the two Works Committee representatives and two members of the Research Team present at every group, one of these latter to take part in the discussion, and the other to record. A short meeting had been arranged between the Research Team members and the two Works Committee representatives half an hour before the first discussion to make sure that everything was set. Unexpectedly, however, the Divisional Manager turned up, explaining that, although he was not expected, he felt compelled to express his fear that the two Works Committee men might so orient the discussions as to have nothing talked about but the size of the pay packet. He hoped that this would not be so, and that they would also discuss in general how to improve the atmosphere in the shop, adding that he was surprised that so little mistrust of management had been reported so far in the Wages Committee.

The Works Committee representatives replied that they had no desire to lead the discussions in any particular way, but wished only to present the facts and to find out what the shop floor was thinking. Accepting this assurance, the Divisional

Manager withdrew, leaving the two trade unionists looking at each other.   One commented "He certainly seems anxious about our discussions".   The other nodded in acquiescence.

The Content of the Group Discussions

The groups ran smoothly.   Each session lasted its full hour, and most had to be broken off to allow the next to come in.   To meet the suspicion present in every group that Research Team members would report secretly to management, the workers' representatives explained that we were there at the request of the Shop Committee.   But in no case did this seem noticeably to diminish the suspicion, and this, the consultant suggested, meant that the workers were suspicious that their own Shop Committee was in league with management.   This interpretation brought into the open such comments as "the Shop Committee are management stooges and buffers", and allowed partial resolution at least of some of the workers' strong concealed doubts regarding the integrity of their committee.

The wide variety of matters raised during the discussions is summarized below, categorized under the rough headings used by the Shop Committee Chairman.

Matters relating to the Pay Packet: Since most people were frustrated by the existing piece-rate systém, the Shop Committee received definite instructions to negotiate with management a change-over to an hourly wage, to be calculated on each individual's existing basic rate plus 65 per cent.   This figure of 65 per cent. was alleged to have been read from a productivity chart posted in the department, which recorded the average weekly piece-rate bonus earned over the past year.   Since the institution of an hourly wage would bring considerable savings on overheads by eliminating the complex administrative set-up required under a piece-rate system, the Shop Committee was asked to secure a proper distribution of the savings throughout the department.

Next came a demand for a slight increase in pay, which grew out of strong expectation that under the new set-up the workers would be able to increase their productivity.   Under piece-rates, every man was out for himself; there was no time to help others out of a jam.   People tried to put aside jobs unlikely to pay well, and these so-called "bad jobs" often led to bottle-necks.   An hourly rate with a secure pay packet would overcome these difficulties, since it would no longer matter if a job was "good" or "bad", and there would be no barriers against co-operative work. And what was more, the workers' wives would also be happier, knowing for sure what money to expect each week.

But what would actually happen if productivity increased? Would management institute an effective and agreed method of recognizing satisfactory efforts? It was proposed that the practice should be introduced of having a continuous measure of the productivity of the shop.

Safeguards and Guarantees: The number of safeguards and guarantees asked for in most groups provided one indication of the morale problems in the shop. The management might wish to change back to piece-rates at a lower level if the new scheme did not work, and the Shop Committee was asked to secure some kind of guarantee that any agreed rate would be upheld. If another department changed over to hourly wages at a higher rate than that which they obtained, would they be able to re-open negotiations to increase their own rate? Or if they settled for too low a rate, this could be used by higher management to drive hard bargains with other piece-rate departments making a change-over.

Under a flat hourly system supervision might begin to "push them around and tie them to their benches". A trial run of the new system for a period of three months was demanded, with the provision that there would be a minimum of interference from supervision; otherwise the workers would never be allowed sufficient time to get used to the new system and show what they could do when they had settled down. The sick, the elderly, and other categories unable to work quickly might find themselves penalized, and how, it was asked, could they ensure that a man would be given satisfactory opportunities to increase his skill and, hence, his basic rate. Appropriate protection against such eventualities had somehow to be obtained.

General Morale Questions. The department was not thought to be as happy a place as it could be. There were mixed feelings toward the supervisors, who were felt to drive the operatives too hard, but mainly because they themselves were being driven by shop management, which "must have something up its sleeve; they must be getting something out of it or they wouldn't have proposed the scheme in the first place". After all, had management not originally put forward the proposal that the operatives should drop a penny an hour on their existing rates, on the grounds that there was likely to be some drop in production if the piece-rate system with its incentives was discontinued? Surely here was proof that management had no confidence whatsoever in them as workers.

A second factor was the memory the workers had of how the original piece-rate system had been introduced by the previous management four years before, when they had been promised that if they did not like the new system, they could discontinue it. In spite of much criticism at the time, nothing had been done about changing it, and the shop was left feeling that the scheme had been forced upon it. As a result, they were fearful that the present management would impose the scheme they were now considering, even if they did not agree to it.

THE INTERDEPENDENCE OF MORALE AND METHODS OF PAYMENT

Shop Committee: 9th March.   The following day the Shop Committee met to consider the results of their discussions with the rest of the workers, and decided to press for basic rates plus 65 per cent., though they would be prepared to come down to 60 per cent. if necessary, with the proviso that before accepting any scheme they must first check the new earnings of each individual so that no outstanding injustices would be done. Although the consultant pointed out to them the strong taboo which was operating against the management and the workers talking to each other about their behaviour, they determined that morale issues, particularly those dealing with the behaviour of supervision and management, were to be held in abeyance for fear that raising them too directly might "get them all thrown out of the office by management".

Wages Committee: 10th March.   A meeting of the full Wages Committee took place the next afternoon at 3 o'clock, at which the Shop Committee put forward their proposal for a 65 per cent. increment. The management, taken aback, explained that the average bonus was 57 per cent., the operatives having made the error of taking the highest figure to be the average. The Shop Committee men were not inclined to recognize their error, and confusion as to where the figure of 65 per cent. had come from led to a stalemate. Apart from one moment when the management expressed appreciation of the Shop Committee Chairman's statement that with a fair wage the operatives would certainly keep up and possibly even increase production, an interminable, if rather polite, wrangling over percentages was all that ensued until the overtly calm and friendly atmosphere turned into silence. This seemed to the consultant an appropriate moment to take up certain aspects of their relationship.

He first pointed out that whatever arrangement they came to they would still be faced with the need to resolve the emotional stresses between the people in the room. These stresses had been demonstrated in the operatives' unwillingness to recognize their error in the discussion of the 65 per cent., and they could be seen again when the Divisional Manager had asked what would happen if productivity went down. The Shop Committee had snapped back that management did not have faith in the workers; and this in turn management vigorously denied.

The consultant then suggested that, if one also took into account what was known from previous meetings—e.g. the Divisional Manager's anxiety on the morning of the group discussions, and management's fears in general that workers could not be trusted to keep production up; the Shop Committee's oft implied suspicion of management; and the first hot-tempered argument between them all, management, supervisors, and workers alike—it was obvious that much more was going on under the surface than would appear from the placid atmosphere in the Committee.

In short, there was evidence both from the present and the past of strained relationships. How could they hope to arrive at a mutually agreed and satisfactory wage level for the shop so long as their fears and suspicions of each other were preventing them from conducting their negotiations with proper effectiveness?

Reactions to these remarks were immediate. The Divisional Manager said that they would be better off if they would take some of these attitudes into account, the Supervisor, that it was about time such problems, which everyone knew existed, were discussed openly, and the Shop Committee Chairman, that they all deserved the "kick in the pants" which had been given them. The consultant observed that it was not a matter of "kicking people in the pants" but that there were issues which affected their discussions and prevented them from reaching agreed solutions.

The atmosphere became easier, and they managed to reach

conclusion on the point that small adjustments should be made in the case of those individuals who stood to lose too much. With this as a kind of successful test-out, one of the Shop Committee members revealed that the shop floor suspicion that management was trying to "fiddle" had arisen from the proposal that a penny an hour should be taken off their wages because of possible loss of production. With such frankness as the keynote, the Divisional Manager replied that he was now satisfied that the shop would maintain production, and was prepared to drop the proposal about the penny deduction.

This constructive atmosphere soon evaporated. Too many other problems were plaguing them. What would happen to individuals who lost money? Could productivity be maintained with the new types of work coming in? Complaints were made about current rate fixing, and there were problems about individuals whose present basic pay was too low. All these questions, though aired, remained unresolved. Everyone was worn out, so that when the Divisional Manager and the Superintendent suggested that they should find some way or other of calculating a reasonable percentage figure and report back, this proposal was jumped at all round.

Two notable features characterized this meeting. One concerned the fact that, hard as they were still trying to confine themselves to the financial aspects of the change-over, they actually spent between 50 and 75 per cent. of the time, directly or indirectly, on the related morale issues. As a result there was an openly expressed dissatisfaction in each meeting that everyone "kept bringing up side issues". The other notable feature was the greater security of the Shop Committee, who now spoke more as representatives and less as individuals, the group discussions having increased their confidence.

Shop Committee: 16th March. Certain morale issues were now becoming pressing in spite of the attempts to keep them back. The Foundry had opened negotiations to change over from a group bonus to an hourly rate, and the Service Department workers' leaders were unsure how their own position would be affected if the Foundry changed over before them to either a higher or a lower equivalent rate. There were also divergent opinions among them regarding the shop's ability to maintain the existing level of productivity—the Chairman arguing that the Committee itself should be able to guarantee a good production rate, while others were afraid that this meant taking too much on themselves, particularly in view of their fear that supervision would adopt a "nose to the grindstone" attitude.

The consultant referred to their difficulty in confining themselves to the wages question, to their lack of agreement with each other, and their anxiety about having too much responsibility, and queried whether some of their suspicion of management did not represent an outlet for their uncertainties about their own position both *vis-à-vis* their own constituents and workers' representatives from other departments.

A heated discussion followed, out of which emerged a general line that was maintained in all their dealings with management during the next few weeks: first they would try to negotiate a satisfactory hourly rate; this having been done, they would then take up morale questions before finally agreeing to the change-over.

Supervisors and Shop Committee: 30th March. At this stage another event of considerable importance in the life of the department occurred. The Superintendent and the Shop Committee Chairman arranged a joint meeting of all Supervisors and Shop Committee members to go into the differences which existed between them. The consultant was asked to attend. Discussion centred on tooling problems, which were used indirectly as a means of talking about their relations with each other. The consultant interpreted the undertone as an argument in which supervisors accused workers of not putting their backs into it, and workers accused supervisors of not properly carrying out their responsibilities. One supervisor thought workers raised these tooling problems just to be awkward. Shop Committee members —and they said this went for the rest of the shop too—believed it was no use bringing these matters up because you got no change out of supervision.

The meeting then got on to the difficulty of timing and assessing piece-rates on some of the jobs coming through. Stating that it was departmental policy to compensate a worker on a poorly timed job by giving him an appropriate *Additional Wages Issue*, the Superintendent looked at the supervisors and said, "That is what you do, isn't it?" As a "yes" came from one of the supervisors, some of the workers' representatives broke in with, "Why don't you ask the chaps in your section what they think about it?"

The net effect of these exchanges was an arrangement for workers' representatives to be brought more into consultation when prices were being fixed. As a member of the Shop Committee put it ,"You don't tell us anything, and expect us to work with you—where is the co-operation round here, anyhow?"

Feeling that the occasion had been profitable, they arranged a further meeting in two weeks' time, and asked the consultant to continue in attendance, as they found it valuable to have "general comments about things which seemed to be going on at the sides of the discussion".

BARGAINING VERSUS WORKING-THROUGH

Wages Committee: 7th April. Whether each side should stick to traditional bargaining techniques, or whether both together should try to work out an agreed rate derived from all the available facts, created the dilemma which dominated the next meeting of the Wages Committee. The Divisional Manager announced that the average percentage bonus of the shop was 57 per cent., equivalent to an average hourly rate of 2s. 11¼d. Then the trouble started. The management referred to the 57 per cent. as the "maximum rate", and the Shop Committee referred to it as the "minimum". The meeting, scheduled to last "five or ten minutes—just time to allow the announcement of the calculated rate"—went on for two and a half hours.

Pointing to the continuing lack of confidence between management and workers, the consultant said that in all their discussions a central issue was the kind of relations they wanted to have in the department. Did they want one power group pitted against another, or did they want to work together? That they were striving to work out something together was clear: but it was also noticeable that from time to time they reverted to bargaining methods; as in the present "maximum-minimum" argument. Bargaining, as traditionally employed, led to the setting up of rates which were the resultant of the relative power of management and workers. In departing from this principle, which in one sense described what they were trying to do, they were facing the problem of what could be considered a reasonable wage for operatives in a factory of this kind at this time. This meant facing sooner or later the complex morale problems in the depart-

88

ment, including the relation between the wages of all groups—management, supervisors, operatives, and office workers.

Without resolving much, they finally arrived at the compromise solution of agreeing to discuss the 57 per cent. with the shop, in order to determine whether the workers wished them to go ahead with further discussions about the conditions under which such a wage could be implemented; that is to say, no commitment was undertaken on either side in taking this figure to the shop floor, and it was realized that they would have to face the possibility of fairly long and serious discussion afterwards.

The First Ballot: 13th May. By the next week the Divisional Manager, the Superintendent, and the Shop Committee Chairman had prepared a document to be circulated to each person in the department. After a summary of the early developments, the document ran as follows:

> *Management started off by considering that the Shop should be ready to accept a slightly lower wage-earning on average because of the advantage of working on a stabilized basis not subject to the hazards of piece-work, and because it was considered that production might suffer to some slight extent when the direct piece-work incentive was withdrawn. Your representatives, however, objected to this view, suggesting that we ought to pay the same amount of wages in the Shop on the new system as on the old, and that it would be up to the Shop itself to give as high an output on the new basis as on the old. Management consider this an extremely responsible attitude to take and accept the principle put forward by your representatives as a basis for discussion.*
>
> *The principle of payment proposed is that each worker should receive a new flat rate made up of his present basic rate, plus a 57 per cent. increment, which is equal to the average bonus earnings of the Shop. This would mean that there would be some levelling out of wages, with less spread between the top and bottom, although an adjustment might be made in the case of a few people where it is considered that injustice would be done by adopting such a basis. . . .*
>
> *The position at the moment is that you are being asked to come to a decision as to whether you would regard the basis outlined above as being satisfactory in principle. There are still important matters to be solved and discussions are continuing on these. They are mainly, the need to preserve a correct relationship between rates earned in the Service Shop and by other Departments, such as Tool*

*Room and Millwrights, and we are proposing to ask representatives
of these shops to discuss this issue with us. Furthermore, we have to
consider to what extent, if at all, there should be any adjustment in
rates established on a new basis in accordance with falling outputs,
or alternatively, increased output, whether this arises from greater or
lesser productivity on the one hand, or greater or lesser volume of
work on the other hand. Subject, however, to final decision on these
issues, we should like to know if you think the general basis proposed
would be satisfactory. On this proposed new basis your own new
rate would be as shown on the bottom of this note. (The actual rate
for each individual was appended.)*

Signed, Divisional Manager
Chairman of Shop Committee.

Following the circulation of this document the Superintendent
and the Shop Committee Chairman together met each person
in the department, offering an individual explanation of the document and inviting anyone with difficulties to see either one of them
separately or both of them together. Six people, worried because
they were likely to take a considerable drop in earnings, came to
see the Shop Committee Chairman, who brought in the Superintendent, and between them certain increases were agreed in
individual cases where real hardship seemed to be occurring.

TAKING UP THE UNDERLYING MORALE PROBLEMS

On the afternoon of 13th May, at a closed meeting of the
workers and their elected representatives, the shop received a
comprehensive report on developments thus far, and voted
unanimously to have their representatives carry on with negotiations for a flat rate. Accordingly, a series of meetings was arranged
between the Personnel Manager, the Works Manager, the Superintendents and elected workers' representatives of the Millwrights
Department and of the Tool Room, representatives of the
Grade III Staff Committee, and the management and workers'
representatives of the Service Department. The relationship
between the rates of the Service Department operatives and those
of the Tool Makers, the Millwrights, and junior supervision was
thought satisfactory. Certain anomalies in individual cases in
the Tool Room and the Millwrights Department were, however,
thrown up. These were taken up by management and workers

together, and the rates of certain individuals in these two departments were increased.

It was interesting to note that because the Service Department negotiations were not treated in a sectional way either by the management or workers, they provided the opportunity for straightening out certain wages problems—and, hence, morale difficulties—in other departments, although this process was not taken as far as it might have been at the time in relation to yet other departments.

Wages Committee: 31st May. With the wage relationship to Tool Room, Millwrights, and Supervision cleared, the Service Department was now in a position to finalize arrangements, and the Wages Committee began to tackle the more general morale questions which they had found so distracting. The main points raised by the operatives during the group discussions in March were summarized on a sheet which the Shop Committee Chairman had kept, and which he took out at this meeting, saying that he and the Shop Committee had a number of points to raise. The sheet ran somewhat as follows:

1. *Pay packet issues*
   (a) *Negotiate on a 65 per cent. basis*
   (b) *Take up who gets the savings on overheads*
   (c) *The rates and adjustment on rates if production goes up.*

2. *Safeguards and Guarantees*
   (a) *Make sure the pay packet agreed will be protected*
   (b) *Get a guarantee of no change back to piece-work at a lower rate*
   (c) *See that Service Department rates will not set a precedent for other departments, and that, if any other departments negotiate a higher rate, Service Department negotiations can be re-opened*
   (d) *Make sure of satisfactory conditions of work with a minimum of interference from supervision*
   (e) *Get security for special individuals such as sick, or older people, or slower workers*
   (f) *Arrange proper facilities for workers to increase their skill.*

3. *General Morale Issues*
   (a) *Get out the causes of the present disharmony in the department*

(b) *Cannot the shop be more independent from the rest of the works, as it used to be?*

(c) *Find out why the section supervisors seem to be more driven and less cordial*

(d) *Take up why the management does not trust the shop to keep up production*

(e) *Find out if management has anything up its sleeve*

(f) *If the workers do not want the scheme, will it be forced upon them?*

(g) *If the scheme is accepted before the summer holidays: will members taking holidays before the scheme is implemented be reimbursed ?*[1]

They went through the above summary point by point, first agreeing 57 per cent. as the new basis on which they were negotiating, subject to final confirmation by the district trade union officials, who had been kept informed of all developments. As regards the eventual disposal of the benefits from savings on overheads, and whether there would be any merit increase if the Shop managed to increase productivity, the Divisional Manager issued the reminder that although Service Department profits were apparently high, this was largely due to the low cost at which replacement bearings were supplied from the main works; and increased profits, therefore, should go to the company as a whole, in whose prosperity the department would share. The Shop Committee Chairman did not argue, but put down "no agreement" beside these points on his sheet.

The atmosphere up to this point had been friendly, but with conditions of work under the new scheme as the topic the conversation became more heated.   The workers complained that other departments always seemed to get the best machines, equipment and everything else, leaving the Service Department as the

[1] All these points except 3(g) arose during the group discussion in March. This last point arose during May, and some explanation is necessary. The workers get two weeks paid holiday each year; the amount of pay being calculated by taking the average of a man's basic rate on 1st July of the preceding year and 30th June of the holiday year. Thus, if the Service Department had changed to a higher rate of basic pay before 30th June, this would have meant a higher rate of pay during the holiday. The practice was for the whole factory to shut down during the two weeks holiday period, with some few individuals taking their holidays earlier and remaining as a kind of skeleton staff during the holiday break. If, however, such individuals in the Service Department took their holidays before the new wage rate came into effect, they would lose any benefits accruing. The above point, 3(g), was meant to cover this particular difficulty. As will be seen shortly the matter of holiday pay became an issue of considerable importance and nearly wrecked the whole scheme.

poor relation; and the Divisional Manager's explanation that the different nature of the work in other departments necessitated new equipment far more frequently was of no avail.

Referring to the increasing heat of the discussion, the consultant interpreted what again seemed to him to be an implicit statement of the workers' lack of confidence in their own management. He commented that the Shop Committee felt that a management "who got talked out of their profits" were stooges to the rest of management in the company. In view of the rational explanations that had been given, he went on to say that many of these feelings represented attitudes towards past managements, which were being projected into the present situation. This point was not taken up, and the meeting turned to consider the course to be adopted if other departments went on to an hourly rate at a higher level than theirs. The Divisional Manager could give no guarantee as to the outcome, and once again the Shop Committee became worried and the Chairman wrote "no agreement" beside this point on his sheet. Asked by the Divisional Manager what he meant by "no agreement", the Shop Committee Chairman replied that these were matters of such importance that the negotiations could not proceed unless some agreement could be assured regarding them. A deadlock ensued.

The consultant asked whether they could agree on any of these points, most of which were matters for the future, without setting up some mechanism for discussing general department policy. Without some mechanism in which management, supervision, and workers alike had confidence, a deadlock such as that existing at present was always likely to occur. The Supervisors' representative expressed agreement, but it was too late to take up so large an issue.

In closing the meeting, the Divisional Manager said he would like time to consider in detail the list of points raised by the shop, and the Shop Committee Chairman gave him the list to be typed and circulated to everyone. The circulation of this list represented a striking change. The morale problems of the shop were now out in the open. Thus, in order to arrive at a wages agreement, the Wages Committee was committed to the task of attempting the satisfactory resolution of these problems.

Supervisors and Shop Committee: 18th May and 1st June. In the meantime, the supervisors and Shop Committee had been getting on with their meetings, and working through a wide range of difficulties. At their meeting on 18th May, following an interpretation of strained relations in the group, one of the workers' representatives took the plunge and brought out the general feeling in the shop that the supervisors were looking more glum and wearing longer faces than they should. This comment was taken in good faith by the supervisors and led to a serious discussion about the difficulties which supervision and shop management were feeling because of the increasingly competitive economic situation.

Then, at the next meeting on 1st June, an event of some significance occurred, when the superintendent, conscientiously trying to provide information on all matters of importance, reported on certain rebuilding plans that were being put under way in the shop. This information, instead of being gratefully received, led to sharp criticism of this expenditure both from supervisors and workers. Apparently there had been a long history, about which the Superintendent knew nothing, of hostility towards the previous departmental management for having engaged repeatedly in expensive rebuilding without consultation either with supervision or workers. The whole of the unresolved bad feelings about these past events flowed out into criticism of the present rebuilding plans.

The consultant observed how their discussion illustrated the inadequacy of merely reporting to people what you had already planned to do. To be serious about consultation meant taking people and their attitudes and feelings into account before final plans were laid; hence, in order to have employed consultation in this case, it would have been necessary to have reported the rebuilding proposals at a much earlier stage, and to have obtained general agreement before going ahead with plans. This interpretation led on to a more general discussion of the principles of consultation and their application to the day-to-day running of the department, a discussion which was contained at subsequent meetings of this group.

ESTABLISHING THE SHOP COUNCIL

Wages Committee: 2nd June: 10.00 a.m. The Wages Committee met again in the morning of 2nd June, just two days after their previous meeting. The atmosphere was tense, everyone sensing that a critical point had been reached, and this attitude became reflected in what turned into a full day of intensely serious work.

The Divisional Manager opened the meeting by stating they had three main points to consider:

1. A guarantee from the shop that production would not suffer.
2. A guarantee from management to the shop concerning benefits from increasing productivity.
3. The establishment of some mechanism for making possible increased participation of the total shop in the making of departmental policy.

The first two points had arisen frequently in their previous discussions, but the third expressed a new attitude on the part of departmental management. In explaining the origin of this attitude he referred to his realization at the last meeting that the workers felt that departmental management were prone to give way to the demands of other departments. He also referred to a report he had received about the furore caused in the meeting between the Supervisors and the Shop Committee by the plans for rebuilding in the shop. He had not expected that those plans would create so much hostility and resentment, since they had been fully discussed with the people in the stores, who, after all, were mainly concerned. He and his colleagues therefore considered that, if they were seriously to go ahead with joint consultation, it would be necessary for them to take supervision and workers more fully into consultation on shop policy.

With these comments from the Divisional Manager, the meeting moved on to a discussion of the list of points which had been circulated, about which, now that they were out in the open, they could speak frankly. The workers asked management point-blank whether or not they had anything up their sleeve—whether they were hiding anything. Management denied this, and made a declaration that they had been quite open throughout the negotiations.

A number of points were quickly dealt with. Special individuals would be taken care of, and opportunities to increase skill arranged for those seeking up-grading; if workers did not want the new scheme, it would not be forced on them; and if the scheme was introduced before the summer holidays, shop employees who took their holidays beforehand would be reimbursed. The more general morale issues, they saw, would

H

take some time to clear up, but now that a start had been made on them, they would go on until solutions were obtained.

Nevertheless a number of awkward questions still remained. Who got the savings on overheads? What happened if productivity went up or down? What happened if any other department got a higher rate? What guarantees were there that the agreed flat rate would be secure, and that there would be no change back to piece-work at a lower level? It was management's view that a policy-making group for the department should be established and these issues referred to it. The Divisional Manager used the words "shop council" to describe such a body, and declared his willingness to vest authority for policy-making in a shop council of this kind, if supervisors and workers would agree to co-operate.

This proposal served once again to bring to the surface the workers' feelings of suspicion and mistrust. They found it difficult, they said, to have confidence in such a plan because of previous experience with management and supervisors. They again referred to the way they considered the "minutes system" had been imposed four years before and to the existing discontent about rate-fixing. The consultant indicated how suspicion and mistrust, arising out of past as well as present experience, had turned into a barrier between workers and that greater stake in the running of the department they had so frequently claimed. The Shop Committee members looked at each other, and their Chairman, trying apparently to gauge the feeling of his co-workers, said they would be willing to look further into the question of what a shop council might do.

The Superintendent suggested that the Shop Committee should meet by itself, talk over their attitude towards the setting up of a shop council, and, if they so wished, bring back specific suggestions on how they thought such a body might be established. This idea was accepted, and arrangements were made for the Shop Committee to meet later that afternoon, while the Shop Committee Chairman, at his request, would meet with the consultant meanwhile, in order to clarify plans to bring before his committee.

Shop Committee Chairman, 2nd June: 1.00 *p.m.* The Shop Committee Chairman was in a dilemma. He had been a staunch trade unionist for thirty years, and carried with him a burning suspicion of management built up through a variety of experience in some of the most severely distressed industrial areas in Britain and Scotland. Mixed with this suspicion was his wish to realize what he expressed as "his dream to see the workers participate in management", not just for himself, but for the younger workers who were now growing up. "My industrial life has been hell, and I don't want to see my children go through the same thing. But it's difficult to take an opportunity like this when you see it, when you've had the kind of experience in industry that I've had in the past."

The consultant talked with him for nearly three hours, during lunch and after, about his conflicting feelings, and gradually he began to clarify his thought and spontaneously outlined a variety of possible ways in which a shop council could be set up, and what it could do. These he crystallized into a proposal that a Council of some twelve members should be established, representative of all sections of the department, and composed of shop management, the full Shop Committee, and representatives of stores, of the clerical staff, and of supervision.

Shop Committee, 2nd June: 4.00 p.m. Following this discussion, the Chairman called together his committee, and put forward his plans. But some of his fellow workers were just as mistrustful of management as he was. As one member put it, "I don't intend to put the next four years of my life into something that can't possibly work out." The consultant stressed that he could not advise them whether or not to co-operate in a shop council with management. That decision they must make for themselves. But it did seem to him that, as a group of workers, they had arrived at a cross-roads, and he felt constrained, therefore, to point out how their own anxieties and suspicions were effectively inhibiting them from arriving at any decision whatever.

They realized that their conflicts were not easily to be resolved. With some hesitation, therefore, they adopted the line that they had nothing much to lose in going ahead. They could give management a trial and, if things did not work out, at least they would be no worse off than they were, and would know better where they stood.

Wages Committee: 2nd June: 4.45 p.m. Accordingly they brought together the other members of the Wages Committee and informed them that they were willing to go ahead with the setting up of the shop council, to which they would refer all outstanding issues that had arisen during the wages negotiations.

The Wages Committee was thus placed in a position to take up with the Shop whether they would finally agree to the change-over on a 57 per cent. basis. But before this was done, final agreement to the wages change-over and to the proposal to establish a Shop Council had to be obtained from the trade union officials and from the Managing Director. The Shop Committee Chairman was delegated to see the former, and the Divisional Manager, accompanied by the consultant, was to see the latter. The agreement of the trade union officials was readily obtained, since they had been kept in touch all along with what was going on. The discussion with the Managing Director was, however, a bit more difficult, since he was anxious lest the Service Department might be trying to set up its own show separated from the rest of the factory.

The Service Divisional Manager, however, put a forceful case to show that the proposed developments in the department were a natural and integral part of the company's policy of joint consultation. Although he was not sure where it would lead, he was most anxious to get ahead and see if consultation carried forward in this way at shop floor level would lead to improved relations. He himself felt it was the correct thing to do, and might have important results.

This provided an opportunity to clarify the role of the Shop Council. It would be responsible for matters which affected the Service Department alone; on questions which affected other sections of the company, however, it would clearly be necessary for management and workers in the Department to consult with the Works Council. This point having been straightened out, the Managing Director agreed the scheme and gave the Service Department full scope to make whatever arrangements were necessary.

Wages Committee: 13th June. The decision to refer to Works Council a matter which affected policy outside the Service Department compelled the workers' representatives to consider their attitude towards the rest of the factory and their doubts as to whether they could operate effectively the social structure outside their own department. These questions arose when the Divisional Manager reported his conversation with the Managing Director, and was attacked by the Shop Committee, who criticized the Works Council and the firm's consultative set-up, maintaining that top management could get whatever it wanted by talking the workers' representatives out of their demands or their arguments.

This clearly expressed fear of management reminded the consultant that it had been at the Shop Committee's suggestion that he had accompanied the Divisional Manager to see the Managing Director the previous week. It would seem, therefore, he said, that the workers now perceived him as someone who could somehow magically protect them against the rest of the works, and particularly the management. If, he argued, this was the case, did it not indicate some insecurity and a fear that they were not strong enough to cope by themselves with their present situation?

This was denied by one of the workers' representatives, who said that the workers did not expect the consultant to solve problems for them, but that they did find an outside interpreter useful. The point was not completely denied, however, for the Shop Committee members had a short discussion among themselves, in which they came to the conclusion that they had to get on and try to solve their problems on their own.

The consultant then went on to say how much both the management and the workers were carrying on tradition inside themselves, in spite of the development which had taken place in the firm. From time to time members of the management had shown anxiety over entrusting authority to their subordinates, and the workers were bound down by their own mistrust and suspicion, and what seemed to be a fear of losing the present situation in which they could criticize management action and yet feel little or no burden of responsibility.

The same worker who had spoken a few moments before said in a rather critical way "What you mean is, we should trust each other and let it go at that!"

The consultant, commenting on how much suspicion was implicit in this remark, referred back to his earlier interpretation. Such feelings of suspicion seemed to be maintained to avoid facing lack of confidence, and from fear of an inability to operate the social structure of the company so as to meet their own needs and the needs of others. He emphasized that the task of finding correct and satisfactory relations among themselves, and with other sections of the company, was clearly one which was still on their agenda. This topic was then dropped, and they decided to go ahead with a final ballot.

IMPLEMENTING THE WAGES CHANGEOVER

The Second Ballot, 16th June. The ballot was held on 16th June. Immediately preceding it, the Superintendent and the Shop Committee Chairman addressed the Department to make sure that all were quite clear about the issue upon which they were voting. The ballot paper read as follows:

> *Do you wish to change the method of calculating your wages from the present piece-rates to a flat hourly rate?*
>
> *If you agree to this change, you must leave to the findings of a Shop Council (to be set up as representative of all London Service Station personnel) further discussions as to what is to be done if production rises or falls as a result of this change. The Council, when constituted, will discuss and decide on all problems arising out of this issue. Such decisions may, of course, need agreeing with the Managing Director and/or Works Council, where Main Works interests are involved.*
>
> *YES, I do wish to change.*
>
> *NO, I do not wish to change.*

In all, forty piece-rate workers in the department voted. Twenty-eight were in favour of the change-over, and twelve were opposed.

Wages Committee: 17th June. The day after the ballot the Wages Committee considered whether or not this 70 per cent. majority was sufficient to warrant a change-over. In order to assist the discussion, the Superintendent provided the following figures. Under the new scheme, seven people in the department would lose from 2d. to 9½d. while thirteen would gain from 2d. to 7d. per hour. Total pay lost per hour would be 5s. 2d., and the total gained 6s. 5d. per hour. In other words, the earnings of the department as a whole would increase slightly under the new arrangement.

The Shop Committee decided that they would be acting in the best interests of the shop by accepting the majority view, but were worried about what management might do if the people who were not in favour of the scheme should for a period of time remain unco-operative. In reply, the Divisional Manager expressed the opinion that if the situation which they feared became such as to require drastic action, he would refer it to the Shop Council.

The Shop Committee became angry, their Chairman protesting "You can't make us responsible for disciplining people. Put your cards on the table. What do you intend to do? Suppose all twelve worked in a half-hearted way, would you sack them instantaneously?"

The Divisional Manager protested he was not "holding any cards", but was being quite straightforward.

The consultant asked whether it was not true that whatever management replied, they would be wrong—for the workers' representatives were once more testing them out at a crucial point in their negotiations. He exaggerated the example, and asked whether the management would dismiss those who had voted against the change-over if they downed tools. Management immediately replied in unison "of course not", and the Shop Committee, somewhat reassured, concurred that it would be a drastic situation and would require discussion jointly between management and workers. This led to some clarification of the potential functions of the Shop Council and of its members. The Council would establish general policy, and full responsibility and authority would be delegated to management to carry out this policy, subject always to checking and criticism by the other members of the Council.

The Wages Committee then adjourned for ten minutes, while the Shop Committee held an independent meeting before coming to a final decision. Immediately the others had gone, the Shop Committee members unleashed the anger which they had held bottled up, for, in their estimation, management had been rather reticent over what they would do about workers who were not fully co-operative under the new scheme. In order to test this assessment, they would attempt to get a written agreement from management not to fire anyone whose work was affected because they were opposed to the change-over.

With this test-out to serve as guarantee, they reaffirmed their decision that it would be for the welfare of the shop as a whole to put the new system into effect, since unless they changed over, those unskilled workers now earning anomalously high wages would be holding up the others. Better relationships could be established with a flat-rate system, in which wage levels were determined more by skill and ability than by the quirks of individual piece-rates.

The rest of the Wages Committee then returned, and the Shop

Committee reported their readiness to implement the change-over, if they could arrive at a final agreement on the question of dismissal. Management had no objection whatsoever to their proposal, and a short note to the effect that there would be no arbitrary dismissals was written into the minute book and signed. Largely through the patience of management, the suspicions of the workers had been dealt with and another test-out situation successfully passed through.

A Report to the Shop: 21st June. At this stage the consultant became concerned whether the Shop had been sufficiently informed of the reason why their Committee had taken the decision to change over. He therefore got in touch with the Shop Committee Chairman and asked him whether it would not be wise to meet all the operatives in the department and give them a detailed explanation of the way the negotiations had been carried on. This the Chairman thought was a useful idea, and, having made the necessary arrangements through the Superintendent, spoke to the workers at a closed meeting on 21st June. His report contained a number of points which indicate how far changes had occurred.

He explained that the Shop Committee had decided there was a sufficient majority for the new scheme on the following grounds: it would allow more equitable payment on the basis of skill; it would overcome the present difficulties in assessing proper piece-rates, and remove many pay anomalies in the department; and it would provide a fair wage for the great majority of the department, and hence would lead to increasing harmony. There was also the value of greater security to the individual, in having a fixed and known pay packet. He then reviewed in detail the way the negotiations had been carried on, and showed the manner in which they had taken up with management the various points raised in the group discussions in the shop. Finally, he stated that the Shop Committee had tested management very severely, making many criticisms and creating a great many difficulties, partly at least in order to see how management would react. On the whole, he felt that management's attitude had been fair throughout and he himself now thought that there was a reasonable hope of obtaining co-operative working relations between the management and the workers. Following the explanation, the Shop supported its Committee's decision and agreed to give the new scheme a fair trial.

The Holiday Incident

The troubles, however, were not yet over. Having decided on the changeover, the workers requested that it be implemented before the holidays, in order to get the advantage in that year of the increased holiday pay which they would receive. This the management readily agreed to do but, on checking with the Finance Office on 25th June, they were chagrined to discover that, unless the change was implemented before 30th June, they would not benefit, since that was the day on which the holiday rates of pay were calculated. And even if they changed before 30th June, they would not get the full benefit of the increase, since their coming holiday pay would be the average of the new rate and the rate in force on 1st July of the previous year. An emergency meeting of the Wages Committee was called and the management explained the position.

The attitude of the Shop Committee was "Ah-ha! so this is what you've had up your sleeves all the time." The management, however, was firm, and pointed out, as was indeed the case, that the workers were just as responsible as anyone else for knowing about holiday wages regulations. There then followed some heated discussion, during which the consultant had an opportunity to indicate how once again the workers were testing management sincerity. Their increased capacity to speak frankly to each other allowed these suspicions to be resolved, and the workers to recognize and admit that they were just as much in the wrong as management. They decided to send the Divisional Manager to find out from the Managing Director whether any special arrangement could be made.

The difficulty was complex. The Foundry had changed over from a group bonus to flat-rates earlier that week, and the Works Director had said they would get their holiday pay on the new rates. He too, realizing his error, had gone back to the Foundry, and the same trouble had arisen there. Hearing about this the Service Department decided to sit tight until it saw what the Foundry would do.

The two departments were told by the Managing Director that they could have the full new rates for the holiday if they felt this was fair. The Foundry decided to take it. The Service Department Shop Committee vacillated. The Superintendent took the Shop Committee Chairman to task for his vacillation, because every hour that went by meant that it was becoming increasingly difficult to get the accounts out by 30th June if they should decide to take the new rates. Pushed into a decision, the Shop Committee followed the Foundry, and took the full rate because everyone in the Shop had expected it, and they wanted to give the new method the best possible start. By a considerable effort the Superintendent, the Shop Accountant and the office staff were able to get the accounts out just in time.

There was general satisfaction in the Shop over management's special efforts, and what seemed at first a nasty situation was handled in such a way that it contributed to better departmental morale and management-worker relations. On 28th June the new method of payment was implemented.

CONCLUSION

What had begun as an issue to do with wages and methods of payment soon led into the complex ramifications of inter-group stresses so frequently tied up with wage questions. These were summed up on the sheet of paper which outlined the pay-packet issues, the safeguards and guarantees, and the general morale problems which were raised in the group discussions with the shop. The seriousness of the managers and of the workers' representatives in their desire to arrive at a constructive solution of their problems allowed them to face and explore this wide range of questions and attitudes which kept cropping up and obtruding in such a way as to hold up their discussions. Their constructive purpose being strong enough to withstand the strain of working-through differences which occasionally reached the point of violent discussion, it became possible for the department to move up to an entirely new plane of discussion, and to accomplish the change-over to a new system of payment by means of the creation of an entirely new institution, a Shop Council, which gave promise of being a mechanism through which members could take part in setting policy for the department.

The process of working-through the wages problem by recognizing and doing justice to the wholeness of the pattern of attitudes, group relations, administrative practices, and technological changes of which the wages question forms an integral part, made it possible for the shop to achieve a double result: they introduced a new method of payment which provided a generally accepted and favoured solution to the immediate problem; but more than this, they have set in motion a process and an institution which will ensure for them that, however the new methods work out, it is likely that they will be able to deal more readily with similar problems in the future by being able to recognize them earlier, and by being better equipped to cope with them as they arise.

CHAPTER FIVE

THE WORKS COUNCIL

Problems of Worker-Management Co-operation

COLLABORATION with the Works Council at Glacier quickly demonstrated how difficult it could be even to establish what seemed like perfectly straightforward changes in organizational methods for improving committee efficiency. The Council, at the time this sub-project began in January, 1949, was composed of eighteen members, a chairman and a minutes secretary. The nine Works Committee members were elected directly from that body, and the nine management members, with the exception of the elected representative from the Superintendents Committee, were appointed by the Managing Director. The Council ordinarily met once a month, immediately after working hours, so that any member of the firm could attend in the Strangers' Gallery. Usually anywhere from ten to thirty people took advantage of this opportunity. The proceedings were formal, but within this formal structure the Chairman judiciously managed to allow as much informality in discussion as possible, and all in all, a very effective balance between the extremes of formality and informality was maintained.

"OUTSTANDING PROBLEMS OF JOINT CONSULTATION ..."

There had been a growing feeling among Works Council members that their meetings were becoming longer and more drawn out, and yet they were not accomplishing as much as they might. The Chairman had frequently voiced the opinion that better preparation for their meetings would lead to more effective discussion and a more rapid and efficient way of dealing with their business. He had occasionally expressed the view that the

Works Council needed a standing committee machinery, and formally proposed this at a meeting on the 24th of January, 1949. This proposal was taken up by some of the management and the Works Committee members, who suggested that there were many issues about joint consultation which remained unresolved, such as the shop stewards question, the problem of Grade III Staff representation, and the cost both in time and personal effort to individuals who worked all day, and spent their evenings in committee.

Partly stimulated by the background study which had been reported by the Research Team to the Council the month before, members of the Council asked whether it might not be a useful idea to have the Project Subcommittee take up the whole question of joint consultative difficulties, enlisting the support of the Research Team in this. A resolution was passed "to ask the Research Team through the Project Subcommittee to help in solving outstanding problems generally relating to joint consultative machinery throughout the company".

The Project Subcommittee and the Research Team, on discussing the resolution from the Works Council, realized that they had become involved in a very complex undertaking. Behind what at first sight seemed a simple problem of improving Works Council procedures, lay concealed a number of unresolved policy questions about the Works Council and its functions, that had been growing for a long period of time, and were making it more and more difficult to get through the business at meetings. Accordingly, they drew up a short report to the Works Council in which they pointed out that whatever starting point was taken, it was likely that a wide range of issues would emerge which would have to be tackled. Then, to be sure that the Works Council really agreed on the intention and scope of the proposed work, a list of examples of such issues was given, including:

(a) *the history and the development and use of consultation in the factory;*

(b) *the relation of formal consultation to the executive authority structure as reflected in the relations between Works Council, Works Committee, Divisional Managers Meeting and Board of Directors;*

(c) *the role of the shop steward (when he is not a Works Committee* representative) *in joint consultation, and the effects of joint consultation on union organization;*

(d) *the structure of the Works Council and the principles of representation;*

(e) *procedures in Works Council, Works Committee, Shop Committees, etc.;*

(f) *the effectiveness of formal and informal channels of communication;*

(g) *methods of training for joint consultation and encouraging increased participation;*

(h) *the value of formal consultation in comparison with the cost to individuals in time, money and personal effort.*

Since not all of these issues could be tackled at once, the Project Subcommittee proposed that a start should be made on the following two problems which seemed at the time to affect people most sharply:

(a) *Consideration of possible procedures for increasing the efficiency of operation of the Works Council, e.g. formation and composition of Standing Committees.*

(b) *An estimation of the value placed on consultation by people in general; including an evaluation of the extent to which joint consultation, both informal and formal, was used, the numbers of people and the time involved, and possibilities for distributing responsibility more widely.*

The first of these, the procedural question, was seen as a short-term problem which could possibly be resolved in a few months and which would help the Works Council carry out its business more efficiently. The second was seen as a longer term research which would gradually lead deeper into the problems of joint consultation and possibly allow them to be taken up on a step-by-step basis. The report was accepted by the Works Council at its meeting on 22nd February, 1949, and the Project Subcommittee was instructed to get on with the job. With this background, the planning group composed of the Project Subcommittee and the Research Team decided on three steps: first, that the Research Team would review the Works Council minutes to determine what had happened to Works Council meetings, how the issues they were discussing had changed, and whether the time of meetings had lengthened; secondly, each member of the Works Council would be interviewed by the Research Team to determine their views about Works Council procedures, and about the time and cost to them of joint consultation; and thirdly, the Works Council Chairman and a Research Team member would draw up proposals about Works Council procedure. For purposes of description here we will treat each of these three steps separately.

SURVEY OF WORKS COUNCIL MINUTES

The Council, when it first began in 1942, had dealt with a fairly large number of matters of principle which had been left unsettled from the past, but then turned in its second year to rather narrower aspects of problems to do with the canteen, holidays, overtime work, lateness and absenteeism. Gradually, however, it began to shift back towards more weighty matters of policy, and this was reflected in a steady yearly increase in the number of decisions taken affecting factory policy, as shown in Table Two. Moreover, after the superficial complaints had been skimmed off, issues such as canteen, promotion, holidays, and so on, discussion on these same items became more comprehensive. At the same time, an analysis of the time spent on Works Council meetings did not uphold the very strong feeling

Year	No. of items
1942	12
1943	9
1944	3
1945	7
1946	12
1947	14
1948	22
1949	31

Table 2: Number of policy items settled each year by Works Council.

of Works Council members that their meetings had become longer and more drawn out. The meetings actually became more frequent between 1942 and 1946, but since then had remained about the same, as illustrated in Diagram Ten.

Diagram 10. Time of Works Council meetings.

The Research Team's interpretation of these findings was that, as the Works Council dealt with more complex and difficult questions, they found it more difficult to obtain satisfying and clean cut solutions. Resolutions became less crisp and comprehensive, and each discussion less well rounded off, so that meetings gave the impression of being longer because they never seemed quite complete.

ATTITUDES OF WORKS COUNCIL MEMBERS

The topics covered in the interviews with Council members included the cost to themselves of joint consultation, their attitudes towards the functions of the Works Council and joint consultation generally, and their attitudes towards the setting up of standing committees for the Works Council. On the question of the cost of joint consultation the following views were expressed.

Management representatives did not suffer financial loss from taking part in consultation, but in all cases there was some incursion into leisure time, either through attendance at meetings in the evenings (e.g. Works Council, special managerial meetings, shop committees, staff committees, etc.) or because meetings during the day necessitated taking work home at night. A large proportion of day-to-day management work consisted in "consultation", in the form of discussions with their colleagues or subordinates and meetings with individuals or groups of people affected by management decisions. At times executive decisions were delayed because higher management members were so involved in such discussions that they could not be got together. Some thought that joint consultation in Glacier imposed certain strains that would not be experienced in other types of organization: as, for example, the need to take other people's opinions so much into account; but in spite of this, consultative methods were preferable because they led to an improved atmosphere and overall efficiency.

On the workers' side, four of the nine representatives were on piece-work (one of these on group bonus). Some always took out A.W.I.s[1] for time spent on joint consultation; others rarely did, feeling that they could make up the time lost, or fearing that their consultative work would lose its spontaneity if each time a person wanted to discuss a problem with them they first applied for an A.W.I. Even so there was trouble; the A.W.I. rarely equalled the bonus rate; frequent interruptions interfered with the rhythm of a job, and made it difficult to recover the speed of working; and when members were absent they were liable to lose high bonus jobs. Estimates of time spent by both piece- and non-piece-workers were up to two hours a day, with even longer periods on special occasions, as was the case during the period when redundancy appeals were so frequent. The actual financial losses incurred varied from 3s. 6d. to 11s. per

[1] A.W.I. Additional Wages Issue: piece-workers who are off production work for any officially sanctioned reason were at this time paid their basic rates plus 33⅓ per cent. This was altered in June, 1949, so that an operative could be paid between 33⅓ per cent. and 50 per cent. bonus.

week. But even apart from financial loss, too much time away from the job might affect the individual's rating with the supervision.

Membership of the Works Council, which meets once a month, entailed membership of the Works Committee, which met twice a month, and a shop committee which met once or twice a month; and there were also special meetings of these or other groups. There was an allowance of one hour a month for shop committee meetings, but there was still a considerable encroachment on out-of-work time, especially as many members attended trade union meetings, social clubs, or evening classes, and already had very little leisure time. This encroachment meant loss of family life and giving up other activities which clashed with Glacier meetings. The amount of time required prevented more people from participating in joint consultation.

Slightly more than half of the members of the Works Council were in favour of having permanent subcommittees. A number of the management representatives were in favour, but advanced varying reasons. Some of them believed that the introduction of formal machinery would speed up their meetings, and cut out unnecessary discussion, which usually arose from insufficient knowledge of the issue; while others were opposed to standing committees but in favour of setting up special committees, whenever necessary, so long as their introduction resulted in a greater spread of the load of joint consultative work. The workers' representatives were doubtful of the value of committees but a few gave qualified approval; some were worried about how such committees could be selected, and feared that they might undermine the authority of the Works Council, while others doubted that there was sufficient skill available among themselves to make such a scheme effective.

Attitudes on the Authority of the Works Council

The majority both of the management and the worker members expressed the view that the Works Council was solely an advisory body, while only a few held that it made policy, and even they agreed that, when there was not unanimity, then the management had the authority to act. Some workers' representatives felt that the Council, with greater worker participation, should definitely become a policy-making body in the future, but even on this some others maintained that there was insufficient knowledge and experience on the workers' side to make it possible.

Joint consultation was unanimously held to have been of great value, but it could be improved. Some found their discussions too prolonged and not as useful as they might be because there was inadequate prior preparation. Neither the Works Committee nor the management side considered that they had, in advance of meetings, sufficiently detailed information on proposals put forward by the other side. Some believed that the time occupied in clarifying items of detail on certain issues was such as to leave too little for the discussion of items of major importance; others that, inasmuch as the Works Council was an airing ground for opinions, time spent even in this way was not wasted. A number of the workers' representatives considered the Works Council discussions to be out of touch with the needs of the shop floor.

The lack of uniformity of view about the functions and responsibilities of the Works Council was also reflected in confusion as to the potential powers of standing committees, if they were set up; and the presence of this confusion in turn suggested that the first job that had to be done was to clear up in people's minds the function of committee machinery. Indeed it appeared to those members of the Research Team who carried out the interviews that the introduction of new procedures at that stage with adequate discussion would have been unlikely to improve the position.

RECOMMENDATION ON STANDING COMMITTEES

The Research Team Project Officer, as consultant, collaborated with the Works Council Chairman in drawing up proposals for establishing a comprehensive organization of standing committees. These proposals eventually became the spark which set off a series of discussions about the functions and responsibilities of the Council. Each Works Council member would be elected to at least one of these standing committees, whose purpose it would be to consider proposals before they came to the Council, interpret them in the light of the Council's main lines of policy, and submit recommendations for the guidance of the parent body. The Council Chairman invoked the principle to which lip service had frequently been paid at Council meetings, that it was not always necessary to have both worker and management representation on every committee. He proposed that membership should be decided by seriously taking into account the experience, interests, and availability of the individual Works Council members. For those who were not yet very skilled, the committees would have an important educational function.

The Chairman and the consultant, after examining the past work of the Council, concluded that probably four to six committees in all would be necessary to cover the Works Council's business. They made no specific recommendations as to how these committees might be organized, except in one case, that of an Internal Development Committee, as they called it, whose purpose would be to carry on a continuous study of the social organization and changing group relationships within the factory, and to make recommendations to the Works Council on alterations in social structure and organizational policy as the need arose. The need for some such function to be carried out had been in the mind of the Council Chairman for a number of years. He had recognized that, while there were specialized technological, research, and sales-planning functions in the firm, social planning and development was scattered among a number of individuals and groups and no one was specifically responsible, despite the large-scale social changes which were being introduced.

Reactions to the Recommendations on Standing Committees

The results of the interviews with the Works Council members, the analysis of the Works Council minutes, and the recommendations for a standing committee structure, were presented to the Council on the 20th of April, 1949, in the form of a detailed report, which was adopted with only a few minor changes. The consultant noted that there was little discussion of the report, despite the considerable doubts, expressed in the individual interviews, about the wisdom of establishing a standing committee structure. If the Council members did not themselves wish to raise these doubts in open meeting, then there was little the consultant felt he could do, and it was only later that these other doubts and fears about joint consultation as a whole came out. The Works Council having adopted the first report *in toto*, the Project Subcommittee drew up a series of recommendations on how to implement the new standing committee structure. These included: the setting up of a Steering Committee —composed of the Chairman and two Joint Secretaries—to take decisions between Council meetings on the less important issues and to co-ordinate the work of standing committees; and the recommendation that the Chairman of the Council should not be a voting member of the Council itself, and that election to the chairmanship should be open to anyone in the factory.

It was recommended that five standing committees should be set up, with every member of the Council to be on one, and a number of examples, culled from previous Council work, were given of the type of subject which could be referred to these committees.

(a) *Conditions of Employment Committee* (three members): to consider such matters as wages, hours, holidays, and holiday pay, overtime, sick pay, pensions, piece-rates, rate fixing.

(b) *Production Efficiency Committee* (three members): to consider such matters as engineering development, scrap, production control, loading, faulty work, effect of new processes.

(c) *Internal Development Committee* (four members): to consider such matters as managerial functions, appeals mechanisms, end of clocking on, representation on Works Council, and to take over the functions of the planning and development of the project carried by the Project Subcommittee.

(d) *Public Relations Committee* (three members): to consider matters related to outside bodies, such as Trade Unions, Engineering Federation, company visitors, publicity about the company, economic affairs, housing, local transport.

(e) *General Purposes Committee* (three members): to consider matters which did not fall under the above headings, such as promotion, selection and training policy, financial policy, welfare, canteen, sales policy.

There then followed certain recommendations on procedure for delegating Works Council members to Committees, and each Council member was asked to consider his allocations of people to committees before the next meeting. During the week prior to the meeting of the Works Council on May 24th, the Research Team, at the request of the Project Subcommittee, went over the report with each Council member. They discovered that the proposals on Works Council procedures were causing disquiet and unrest, not so much about procedures as about the role of the Council. A number of the workers were concerned whether or not the proposed changes would increase the status and authority of the Council; some of them hoped this would be so; others hoped it would not, since it would increase the responsibility of the workers, and they felt they had too much as it was. Some members were worried lest the Standing Committees should take over the work of the Council and leave nothing for the parent body to do. As one worker put it, "If I found myself alone on a committee with two other management members, they would be able to argue me out of anything, and take decisions which the Council wouldn't be able to do anything about". Still others feared that the proposed Steering Committee would have too much power or that the proposed standing committees would add immeasurably to the task by creating a number of new meetings to be attended.

These fears had gained currency because the report from the Project Subcommittee contained a list of names of the Council members for convenience in nominating to the various committees. The fact that each member saw his own name listed emphasized the fact that he himself was one of those who would have to become either the chairman or a member of one of the committees. Moreover, they had to place each one of the Council members on a committee, omitting no one, a procedure which precluded the use of their usual device of quickly nominating and electing the same small core of individuals over and over again to committees whenever the need arose, without much consideration of the person's merits or availability for the task. But now they would have to sort themselves out to make up five well balanced committees, and some Council members did not have confidence in the competence of all of their colleagues. This last point of confidence was openly raised at a special meeting held by the Works Committee members of the Council immediately prior to the Council meeting. Had the workers the right to accept the kind of responsibilities implied in the new set-up, and had they the necessary qualifications and the ability to play an active part? No one could give a satisfactory answer.

The Works Council and the Making of Policy

The report on standing committees was placed before the Council at its meeting on May 24th. After it had been read through by the Chairman of the Project Subcommittee, the Managing Director asked the Council if it would give him permission to make a fairly long statement. The report had given him cause for a great deal of thought, and he would like to state openly the nature of these thoughts. The Council agreed, and settled back to listen.

The document, the Managing Director said, needed careful consideration. For some years now the management had been moving along the path of sharing responsibility to a greater and greater extent. Considered in this light, the proposals from the Project Subcommittee could be thought of in three different ways. The first was that the Council should now immediately become the overall policy-making body for the company, so that every major policy decision would become a matter for consultation with the entire Works Council. Secondly, the proposals might be seen as having the limited objective of setting up procedures which would make the existing work of the Council easier, but leave its functions and responsibilities untouched. The third way, however, and this in his view was the correct one, was to say that the Works Council was setting up machinery which would be capable of dealing with all matters of policy for the company, but to realize that at the present time the Council was probably not capable of carrying out this job completely. Therefore, he suggested, the extent to which general policy matters should be considered by the Council would be merely a question of growth and experience.

He then went on to outline two stages in management: policy-making and executive action. The Works Council, he thought, should be responsible for the first stage, and the management completely responsible for the second. The authority of the management in executing policy would thus be derived from the fact that the policy which it was implementing was arrived at through unanimous agreement in consultation with workers' representatives. Finally, he pointed out, if the Works Council were to take an active part in creating policy, it was likely that the standing committees equally would be concerned with policy matters. As a result, therefore, he proposed that these standing committees should comprise both management and workers' representatives.

The Chairman pointed out that the Managing Director's last proposal would mean an amendment not to the present report but to the one they had passed the month before. This information from their Chairman caused the Council members to realize that they really were not very clear on just what they had passed at their last meeting. Anxiousness about what they might unknowingly have committed themselves to increased, and one worker asked when these standing committees they had set up would meet. But when the management in turn asked whether the real worry was that pressure would be brought by supervisors to get them to hold their meetings outside working hours, the workers revealed the dilemma which existed between carrying out their jobs and attending so many meetings, as well as the conflict between meetings after working hours and home life. There were differences of opinion among the workers, some feeling that production work should always come first and meetings second, with no meetings during working hours, while others held that consultation, if it were to be taken seriously, should take place as part of the job, during working time.

One of the Works Committee men then turned to the question of the function of the Council. He did not agree fully with the Managing Director's earlier comments. In his estimation, the Project Subcommittee's recommendations were intended not to extend the functions of the Council, but only to streamline its work under its existing terms of reference. In reply, the Managing Director pointed out that the new set-up could make possible a gradual change-over if they wished it. But another worker maintained that the functions of the Works Council Steering Committee were also involved, because who was to say how the "trivial decisions" which the Steering Committtee might take between meetings could be defined. The Chairman explained that decisions made by the Steering Committee would not be valid until they were ratified by the Council. The Commercial Director, commenting that committees were meeting every day and taking decisions, maintained that experience had shown that these committees were responsible and kept within their terms of reference, a view that was reinforced by others, who held that decisions could always be revoked by the Works Council. One of the workers, however, interrupted to say that the die might have been cast and a decision irrevocable. But his viewpoint was in turn refuted by the Managing Director, who argued that while the new situation might not be perfect, it should be remembered that the workers would be represented on these committees.

The Research Team consultant, who had been co-opted to the meeting for the discussion of this one item, pointed to the considerable anxiety over the possibility of small committees making decisions that would be binding on the Council as a whole. Was this fear of delegating responsibility not a reflection of low morale? There was the frequent complaint that the management could talk the workers out of what they wanted, and hence get its own way at Council meetings. Perhaps this kind of morale problem was the source of one of the anxieties of the workers that if one of them found himself alone on a committee with management representatives he might be unable to hold his own end up.

As though to escape from having anything to do with the awkward points raised by the consultant, the meeting turned to consider methods of electing members to standing committees. But the Managing Director, insisting that it was a very complicated subject they were discussing, argued that they should adopt the document in principle before considering methods of nomination to the committees. His suggestion created an impasse, because such agreement as he asked for just did not exist.

To resolve the predicament, one member moved that the document should be remitted back to the Project Subcommittee for amendment in the light of the discussion which had taken place. But they were all somewhat taken aback when the Chairman informed them that the only amendments that had been proposed were amendments not to the document before them but to the one they had already passed at their meeting the month before. Realizing that what they really wanted was more time to sort out their feelings and ideas about the unexpected problems thrown up by the standing committee proposals, the Council members decided to disregard the fact that they had passed one document at their previous meeting, and referred both of them back to the Project Subcommittee to be modified and amalgamated. They could then be brought forward to their next meeting, and the whole matter considered afresh.

ANXIETIES UNDERLYING THE DISCUSSION ON STANDING COMMITTEES

Were these difficulties, it might be asked, related directly to anxiety about having standing committees? There is much evidence to suggest that this was not the case. The Works Council had always used committees to consider much of its business, and there was little reason to suppose that there would necessarily have been much increase in the amount of time required. It seemed to the Research Team at the time that what had happened was that the two reports to the Council and the interviews with its members had given them all cause to consider for the first time in a number of years what they were trying to accomplish. This had caused many of the frustrations arising from the way the Works Council was operating to be brought to the surface, including the thorny but unresolved issue of the responsibilities and authority to be vested in consultative bodies. On 24th June, the amalgamation of the previous two reports was presented to the Council. Once again a long and involved discussion took place, with little agreement. The fears of the workers' representatives that the standing committees would demand more, rather than less, time had been intensified by the resignation during the previous week of two Works Committee members on the grounds that they were overburdened and lost too much in earnings. The Research Team consultant observed that their deliberations were concerned not so much with the standing committees as with the functions of the Council, notwithstanding the fact that the purpose of the committees recommended by their Project Subcommittee was simply to assist the Council to get its job done more effectively. Their fears about the amount of time they would have to put in surely faced them with the more fundamental issue of how strongly they desired to maintain a consultative set-up.

To take the question of time involvement itself, the consultant said he had noted that they often mixed what were properly matters for joint consultation with what were properly matters for executive action. Was the management not passing on to the Works Council too much of its own work? It had been mentioned that one of the duties of the standing committees would be to get out facts and figures for Works Council discussions. Was this not properly a function to be carried out by the executive at the request of the Works Council? If they allowed the management's executive responsibilities to become mixed up with consultative activities, then the Works Council must inevitably suffer from overburdening. They might avoid such a situation if the responsibility were placed squarely on management to do all the necessary work in connection with the drawing up of rough drafts of resolutions to put before the Council which could then consider them and modify them as necessary. The burden of the work load would thus rest on the particular manager within whose sphere of responsibility a given problem lay, while the standing committees would give prior consideration to matters coming up on the Council agenda so that the Council could have a considered opinion on these matters from a body of its own members.

The Chairman informed the Council that the Research Team was in process of preparing a detailed report to be submitted to them at their next meeting, and that this report would include among other things a more detailed statement of the remarks they had just heard. In view of this impending report, a number of Council members strongly urged, and the rest agreed, that the whole matter of standing committees be kept in abeyance until they had had this longer report.

A Special Report to the Council, 27 July

In its special report to the Council on 27th July, 1949, the Research Team observed that the Works Council had reached a point where the course to be set for the future was obscure. A reconsideration of the Council's function, and particularly its authority and responsibility, was called for. A number of symptoms could be cited: concern about the demands on people's time; feelings that being an elected representative lowered rather than raised prestige; fear of victimization or of upsetting other people if you said what you felt; the recurrence of the same problems year after year at Works Conferences, and nothing done about them; reporting the good features of Glacier to others as an escape from dealing with internal problems.

Contributing to the causes of these symptoms were a number of problems of group relations. Higher management sought to avoid executive responsibility, with the hope that somehow the use of consultation would eliminate the need to give orders; there was the failure to work out effective face-to-face relations between persons and groups in the executive and consultative channels; and was it not just in the area of resolving group problems that there existed such strong suspicions and lack of confidence among people as to hold up effective action?

Problems of this kind could not be resolved without identifying and coping with the attitudes and resistances which prevented their solution. Among these were despair—a feeling that it was no use saying what one felt because no good would result. Anxiety about redundancy was contributing to this, for, although the redundancy problem was handled as constructively as possible, the devastating effects on people in the employ of the firm had only been side-stepped. Moreover, nearly all of the Council members, managers and workers alike, believed privately that the Works Council was not doing as effective a job as it might. But surely, if they did not take up fundamental policy questions which deeply touched and affected people at every level, then they must expect that they would lose pride in their consultative work and feel isolated from the people they represented, as well as meet apathy and dis-interest in others.

There was little reason to believe that they were less affected than other factories by the traditional and deeply ingrained industrial pattern of the workers being suspicious of the management, and the management being jealous of their authority. Not only had such attitudes not disappeared, but it seemed as though the only time the members of the Council could hold their heads high was if there was a management-worker fight on; if there was no fight they felt guilty, as if they were not doing what was expected of them. These management-worker differences, however, were not the whole story. They concealed other and more disruptive differences among the management and among the workers, which, in order to give the appearance of unity, were not exposed in public, but it was these differences which most interfered with satisfactory policy clarification.

The report from the Research Team, of which the above is a summary, was not discussed the evening it was presented, because of a general feeling in the Council that it required more serious and detailed consideration than was possible in the short time available. A proposal was adopted that they should go away *en bloc* for a whole week-end to discuss it, and the first week-end in September was scheduled for this purpose. The meeting then returned to the business of creating a standing committee structure which had been put in abeyance at the previous meeting in June. But now that there was to be an opportunity for a full-scale discussion of some of the problems of consultation which had been worrying them, it was unnecessary for the Council members to project all their fears into the discussion of the standing committees. Accordingly the committees were elected with a great deal of efficiency mixed with good humour in a matter of about twenty minutes.

A WEEK-END CONFERENCE

The Council retired to a training centre near London from Friday, 2nd to Sunday, 4th September, to discuss the Research Team's report under five headings: policy questions requiring clarification; the cost and time of joint consultation; unresolved attitudes; functions and responsibilities of the consultative system; and functions and responsibilities of the executive system.

On the Friday evening there was an ordinary business meeting, which they hoped would not last very long so that they could get on to the main business of the week-end. This hope was not realized. They were trying to decide a date for the two-week holiday shut-down period for the following year, an issue on which there had been a deadlock for two months, with top management in favour of August and the Works Committee in favour of July. They were again unable to resolve the holiday question, however, because some Council members, adopting the attitude of delegates rather than representatives, would not modify their views without reference back to their constituents.

Also on the Friday evening the Council re-elected the Medical Officer to be its Chairman. The election, in contrast with the year before, was not an automatic affair. The Medical Officer stated that he was not keen on re-election, since the chairmanship was not a particularly happy role for him because of the Council's inability to conduct its affairs in a businesslike manner. Taking due notice of the significance of their chairman's comments, and in spite of the reservations that he had stated, the Council members decided that he possessed skills and constitutional knowledge which they required, and asked him to continue, in the hope that the new standing committee structure and the clarification of functions which they were beginning with the week-end meeting would make the Council the kind of body in which the role of chairman would be more satisfying.

The Friday evening meeting was not an auspicious introduction to the week-end conference.   There had been much dogged and drawn-out argument, so that they had only got through the above two items, and had not been able to begin their more general discussions.   Tempers were rather frayed. Many seriously doubted whether the week-end would prove anything more than a nuisance, and even the spirits of the optimistic were damped.   The fear that the Council might be incapable of constructive accomplishment had broken through the attitude of satisfaction with progress which had been used to conceal it.

The Functions and Authority of the Council

It was in the gloom from the evening before that the discussion of the Research Team's report began on Saturday morning. The plan was to cover the five main themes in three sessions, two during the morning and one in the afternoon.   In the evening, the five newly constituted standing committees were to meet, each one covering one of the five themes and bringing back the results of their deliberations and any recommendations they might have to a meeting of the full Council on the Sunday morning.

The discussion during the first meeting on Saturday morning quickly centred on the difficult question of the functions and authority of the Works Council, and this became the central theme of the week-end.   It was started by the Managing Director who had made some notes on the purpose of the company, and on the functions of the Council, the executive, and himself.   The purpose of the company he considered "was the continuity and expansion of a working · community which aimed at the establishment of such internal conditions as would enable its members to serve society, to serve their own dependants, and to serve each other, and to do these things with a sense of creative satisfaction".   This purpose could best be accomplished by "concentrating on the subsidiary aims of seeking the maximum technological efficiency and the greatest organizational efficiency; by seeking to establish an increasingly democratic government of the factory which would award fair responsibilities, rights, and opportunities to the producers, executives, consumers, and shareholders, and by seeking through work to earn such revenue as would enable the company to achieve those objects mentioned under the Financial Policy of the Principles of Organization".

The function of the Council he proposed was "to carry the responsibility of deciding in the light of opinions of producers and managers, and in the light of the interest of consumers and the nation at large, the principles and policies which should govern the executive management of the company. Council delegated to the management, through the General Manager, the job of running the factories in accordance with these principles and policies". Parallel with this, he suggested, the function of the executive should be "to undertake the management of the factories within the framework of the principles and policies, as agreed by Works Council, and in the light of the financial responsibility of the Board, as represented to them by the Managing Director. The executive would bear the responsibility of raising with Council, and making proposals to Council on, any matters upon which there appeared to be a lack of Council policy. In an emergency, executive action might have to be taken which was not within the terms of agreed Council policy —in which case, executives would undertake to obtain retrospective discussion about the policy involved by their action."

The Managing Director's statement led to a spirited discussion of the difficulties of workers taking part in policy making; and it was not simply a matter of workers demanding more power and management denying it. The differences among managers and among workers were as great as those between managers and workers; some management members believed the policy-making authority of the Council should be extended, others that the role of the Board of Directors had to be clarified first; some workers held that they should go forward and accept increased responsibilities for policy making, while others maintained that they already had too much responsibility and feared being saddled with more.

Most of the exhortation for the workers' representatives to assume greater responsibility came from management. In response to this urging, and because of their differences in opinion about the degree of participation they should undertake, the Works Committee members of the Council held a short special meeting on the Saturday evening. Most of them despaired of being able to face any further responsibility, but two or three members, including the Service Department representative who only three months before had faced the same problem in his own department, put forward the case that they must go-ahead if they were to realize their own and other workers' ambitions to take part in making the decisions which affected them. But such a spurring on had little effect on the others who were preoccupied by the criticisms already levelled at them for their inability to present the views of their constituents and for being talked out of things by management. Nor did inspirational argument get rid of the fact that carrying out production work at a machine is not so easily made compatible with committee and other responsibilities during working hours. As factory operatives they were in a literal sense "tied to their machines"— by supervision, by the demands of the production process, and by the criticisms of fellow workers if they were off the job too much.

A very heated discussion on these lines by the workers' representatives threatened to go on for a long time without reaching any definite conclusion. A hasty stop-gap decision was therefore taken to go ahead in a step-by-step manner, accepting added responsibility for policy making in the Works Council as their confidence in being able to cope increased. Those affairs which they did not wish to tackle could, as at present, remain the responsibility of the management.

The Council members got some way towards clarifying the functions, power, and authority of their Council, even though the problem was not completely resolved. They recognized that they had been behaving as though they were representative of the whole concern instead of just the London factory, and they saw that under the existing company structure only the Board of Directors and the Managing Director could be said to hold responsibility for the total concern. To have joint consultation at company level, they would have to create either a new body or a co-ordinating mechanism to bring together representatives of all the units making up the firm. That such a mechanism did not exist was a handicap, but did not prevent the decision being taken that, within the framework of company policy, the London Works Council should establish the policy which would govern in principle the behaviour of managers and workers in the London factory.

The question whether Council members were delegates or representatives was considered, and was seen to be tied up with Council morale. When there was suspicion and insecurity in their relations with each other they took refuge in the more easily determined and more rigidly defined role of delegates, and when morale improved they could accept the more demanding and difficult role of representative.[1]

The problem of the relative importance of committee work and production work was skirted. The consultant's impression was that there was little agreement on just how important joint consultation was for efficient production in the factory. While the most representative view was probably the modest one expressed by one member who suggested that joint consultation was useful for production, there were some who held that in a crisis, consultation interfered with production and should be stopped, and some, at the other extreme, who equally strongly maintained that factory efficiency had improved as a result of consultative work, and it was thus essential, particularly in times of crisis. Factual confirmation, however, was not available to support any of these views. With regard to the skills required for consultative work, the main emphasis centred on workers' representatives gaining some knowledge about the work of higher management. They shunned the more immediate problem facing representatives, how to acquire skill in identifying and presenting the things about which their constituents were concerned.

RECOMMENDATIONS FROM THE CONFERENCE

On the Sunday morning the Standing Committees reported the results of their discussion the night before. The reports were lengthy and the recommendations numerous and constructive. As each report was given, the morale of the Council visibly improved, and by the time the last report was read most of the gloom of the Friday evening and Saturday had been dispelled. The members of the Council, for the moment at least, were reassured about their capacity as a group. Because of the large

---

[1] This was demonstrated at the following meeting of the Works Council in September when, as a result of the temporary heightening of morale achieved during the week-end meeting, the Council members allowed themselves to behave as representatives, and worked out an original arrangement for the summer holiday shut-down, thereby resolving the impasse of the previous few months.

number of recommendations, they decided not to take action on them at the week-end meeting other than to delegate them to appropriate bodies to study and to report back on at subsequent Council meetings. A summary of a few of these recommendations will give some picture of the work done:

To define the functions of Works Council and Shop Councils. This was referred to the Personnel Manager, who was asked to bring back a draft statement to the Council through the Divisional Managers Meeting. As a rough formulation the Committee that made this recommendation presented the following proposal:

"The Works Council is to debate any matters it chooses to discuss; to formulate policy on any matter which clearly does not impinge on or run contrary to the functions of the directors as financial trustees for the shareholders; to delegate, through the General Manager of the factory in the executive structure, the running of the London Factories within the terms of the policy agreed.

"To agree where it so chooses that as a matter of Council policy, certain fields of policy making are, for the time being at least, outside the capacity of the individual members of Council to debate (because of lack of knowledge, experience, etc.) and such matters should, therefore, be left to the unilateral prerogative of management."

To define the functions of the London and Scottish Factory Works Council and the integration between the two. The Managing Director was delegated to discuss this with the General Manager of the Scottish factory. It has led to the Managing Director taking up with the Board of Directors the suggestion that a Company Council representative of the company as a whole might be set up, a matter which is under consideration.

The work of Works Committee members should assume greater importance in relation to their normal work. This was referred to the Works Manager, who circulated a statement to the factory, endorsed by the Council, that committee work was equally as important as production work, and that this should be taken into account by supervision in releasing committee members for consultative duties.

Opportunities should be given to members to see the work done by executives, to allow them to have a wider knowledge of what is taking place. This was referred to the Personnel Director, who arranged for his training staff to discuss with Works Committee the possibility of arranging whatever kind of course they desired. The results of this will be dealt with in Chapter Six.

Works Committee members to consider themselves representatives of their constituents and not delegates when sitting on Council. This was referred to the Works Committee and reaffirmed by that body. The problem was not in agreeing the role, but in behaving as representatives without regression to the protection of delegate status when the going was difficult.

A review or report on Works Council proceedings—similar in form to the Annual Report made to shareholders—should be made once every year. This proposal was adopted and was referred to the Steering Committee for action, along with the instruction that a digest should be made of all policy matters agreed by the Works Council since its inauguration.

Consideration should be given to the laying down of procedures for making good wages lost during works hours, and to payment for time spent out of works hours, for work on joint consultation. This was referred to the General Purposes Committee of the Council, which prepared detailed proposals, now agreed by Council. These proposals include a small annual payment for time spent at meetings after working hours, and payment for time lost during working hours based on the individual's average earnings rather than a fixed sum. The new proposals eliminated a long festering problem.

A STUDY OF COMMUNICATIONS

It finally remains to consider the effects of one other recommendation made at the Works Council week-end conference, that "consideration should be given to the breakdown in passing information to the shop floor". This recommendation, made independently by three of the five standing committees, reflected some concern that the results of Works Council business were not widely known in the factory, and expressed the degree to which most Council members felt isolated from the rest of the factory. The problem was passed to the Internal Development Committee to investigate with the help of the Research Team. The Research Team agreed to collaborate on condition that the study commenced with an assessment of the efficiency of the Council members themselves in communicating with their constituents and subordinates, a condition directed towards maintaining the principle of working only with the agreement of the people concerned.

The plan of the study was twofold:

1. an investigation of the relations of the Works Council members with their constituents or subordinates, in order to ascertain how effective were the councillors in maintaining two-way communication;
2. an examination of the structure of the Council, in order to determine whether structural changes might contribute to increasing the efficiency of communications.

The mechanisms of communication were left out of consideration, because the bulletin boards with their frequent notices, the weekly broadcast of factory news, and the monthly factory magazine were more than adequate in comparison with the possible difficulties arising from organizational inconsistency or the quality of individual and group relationships.

We shall now consider in turn the results of the study of the communications efficiency of the Works Council members, and of the structure of the Council.

The Communications Study—Method

Experience of the disappointing results which have followed the many attempts in industry to improve communications by designing better information services and developing more elaborate consultative machinery suggested that, in studying problems of communication, perhaps too little attention had been paid either to the relations of those attempting to improve their communications or to the effect of attitudes, both conscious and unconscious, upon the quality of communications. The content of the communications study was therefore widened so as to include not only such phenomena as the passing of information, giving of orders, airing of grievances, and the expression of opinion, but also the factors, both conscious and unconscious, which influence communications by affecting people's understanding, feelings and behaviour.

The method used in the study was as follows. Each member of the Council was interviewed; afterwards a discussion was held between him and his next level constituents or subordinates to give these others the opportunity, if they wished for it, to comment on the effectiveness of the particular Works Council member as a link in the communications chain to and from the Council. Then, if the group so desired, all were interviewed separately; it having been made clear that the material from the interviews would be used to prepare a report for the group as a whole to discuss. As a next step, the subordinates or constituents were themselves offered the opportunity for similar discussions with their own subordinates or constituents, and the procedure was thus repeated, step by step.

By this means work has been done with nearly all members of the Finance Division and the Service Division, the top levels of the Personnel Division and Works Division, two production shops, and part of the Commercial Division, comprising some 300 interviews and 105 group discussions, carried out over a period of ten months. Added to the information about the communications system derived in this way, there was direct observation of the process of communication by means of regular attendance at Works Council, Divisional Managers Meeting, Superintendents Committee, Works Committee, meetings of three different levels in the executive chain in the Finance Division, and collaboration with three shops.

The Communications Study—Findings

The interviews with the Works Council members, which began the communications study in November, 1949, showed that most of them had forgotten that they had been at all concerned about the problem. On being reminded that it was their own decision at the week-end conference to look into communications, most of them hesitated and then mentioned that while they themselves had no problem of two-way communications they did believe that there might be some other members who did have such difficulties; but perhaps, after all, it was not really such a serious problem as to require investigation.

This seeming change from intense worry to lack of concern about communications which had occurred during the two months since the week-end conference was interpreted to the Council as a manifestation of their sense of isolation from the rest of the factory. When attending meetings of the Council, particularly when completely away from the factory as had been the case at their week-end conference, they feared that the rest of the factory had little interest in them or in what they were doing, and hence they felt distressingly isolated. During the current interviews, however, they were seen on the job; and in their "real" roles, as distinct from their roles as Council members, they felt both more secure and less isolated, and the problem of communications did not have the same feeling of urgency and sharpness. In short, their outlook at any particular moment was affected by the role they were carrying, and it seemed to be in the Councillor role that they suffered most discomfort.

The fears of the Council members that they were out of contact with their constituents were shown by the communications study to be far from groundless. Very few of the people interviewed felt intimately connected with the consultative machinery in the sense that they wanted to do anything about the way it functioned. The operatives in a shop usually picked one person—he might be their Shop Committee chairman or shop steward—to whom personal grievances were taken, and from whom advice was sought. The rest of the shop committee served as a kind of supportive structure for these persons, who formed the hub of the individual appeal system, but only rarely was the committee used as a means of taking up a problem. Similarly, the staff members did not perceive their staff committees as channels for raising either personal or general grievances, or questions of policy; such things were matters between a person and his immediate superior. On the whole the consultative machinery was not found to be functioning as a mechanism by which shop and office attitudes were collected and integrated so as to affect the decisions of policy-making bodies.

Although people did not keep closely in touch with what went on, or even take elections very seriously, the consultative set-up was nevertheless regarded as an important possession. Its existence was an indication of goodwill higher up, and a mechanism for righting things "just in case" anything went wrong. But on the whole the appeals system worked efficiently, and satisfaction could usually be obtained without bringing the consultative machinery into play as a grievance channel. And apart from personal grievances, the individual found it difficult to perceive how to use the consultative machinery for taking up more general questions. If general problems did exist, it was surely the responsibility of their elected representatives and of the management to recognize and do something about them.

Concern with the Executive Machinery

Those interviewed displayed far greater spontaneous interest in talking about relationships and communications in the executive system than in the consultative system. Thus, for example, there was said to be a nearly complete failure by the management to pass on information about what took place at Works Council. Top management and the workers' representatives took decisions which affected factory policy, but the rest of the management and employees did not often hear officially about the changes, although there might be a notice on the bulletin board. One instance of this was the Council's decision to change the method of paying elected representatives for time spent in consultation. No comprehensive management instruction covering all divisions was ever issued to implement the change, it being left to the initiative of each individual manager. In consequence, many of those who had to implement the change only heard about it fortuitously.

This greater concern of people with the executive machinery did not mean that the consultative channel was seen as operating smoothly. But people felt their day-to-day lives were mainly affected by the way the executive system worked, particularly the quality of their relations with their own superior as well as with the executive next higher. The most frequent complaint about the executive system was that although orders and instructions about work travelled easily enough, it was difficult to take up ordinary feelings, especially if they were critical, about your job or about life in the factory. The main stumbling block in the way of getting such feelings resolved was the reticence about communicating them upwards. This reticence was said to be due to the fact that if a person tried to express to his superior his feelings about his job, or about the superior himself, it was all too likely that the superior would argue with him and try to show him that his feelings were unreasonable and that they did not tally with the facts. Having the existence of one's feelings denied in this way only made things worse. The person was not only left with the original feeling but had in addition a resentment against his superior for not understanding him and not helping him to get at just what it was that was disturbing him.

Barriers to Executive Communications

It was the Research Team's impression that the personnel

from top to bottom of the executive chain had a strong and positive desire to avoid receiving criticisms from immediate subordinates. The attitude was widespread that to avoid formal discussion of personal feelings and criticisms of this kind was justifiable. For how was it possible for a manager to talk over with his subordinates their attitudes towards himself, or even their attitudes towards each other, without becoming involved in considerations of status and prestige, rivalries about position, personal characteristics, and other highly personal matters which it was not really proper to talk about? In part, since many of the roots of such problems lay outside the factory in the private lives and personalities of individuals, the desire to avoid getting involved in them was understandable.

The barriers preventing managers and their subordinates directly resolving their feelings towards each other, in turn created other barriers to communications. For with barriers to personal feelings there could be little dynamic pull up the executive chain to encourage the upward flow of other important attitudes in a formal and recognized way. One way of skirting the problem was to use accessory channels—such as a functional management line or consultative channel—to get information upwards when no results had been obtained by using the appropriate executive line. At times the accessory routes were used in preference to, and without even trying to use, the direct executive channel.

Subordinates would get criticisms off their chests to their own superior about the executives higher up the line, but would refrain from criticism of their own superior himself. The superior might then in his turn take the criticisms up to the next level, but would couch them in terms of criticism directed still higher up. In this way harmony was maintained between each executive and his immediate subordinates, but constructive criticisms could seldom reach their mark. Conversely, executives always approached with great anxiety the time of the year when they had to make out progress reports. Criticizing their subordinates was an uncomfortable task, when at the same time they feared to receive criticisms in return and were unable to criticize their own superiors.

But the difficulty was that unless some means was found to deal with the effects of stresses inside the factory, then the feelings engendered would become mixed up with seemingly unrelated questions about work and organization so that problems which otherwise would be capable of solution became unnecessarily difficult or intractable. To take one instance, a situation arose where top management became worried about an increase in the scrap rate. They passed an instruction down the executive chain that something had to be done about it. At the same time there were current a number of strongly held views in various regions of the works as to why the scrap rate was going up: for example, it was held to be the result of anxiety about the taking on of new workers (because of the fear of a possible recurrence of redundancy), and irritation that people were being pushed too hard by supervision to deal with a heavy load of work.

Regardless of whether or not such factors actually were influencing the scrap rate, the people who held these beliefs were in no receptive frame of mind for instructions about reducing scrap, unless their own views were asked for in turn. In the absence of a strong demand for information to pass upwards, "the management" became the object of criticism for not really finding out what was going on before setting about to right things.

The continuous reporting back to groups, which took place as an integral part of the process of development of the communications study from level to level, made a small contribution towards easing the problem. Supplementing these reports to small groups, there have been more comprehensive reports made to whole departments and divisions, and to the Works Council. Examples of the types of report made will be found in subsequent chapters dealing with the top management group and with the Works Committee. But perhaps the most significant contribution to the improvement of communications has been the gradual clarification of the definition of the place of executive work and consultative work in the factory, features of which have already been described, and which will be elaborated, along with the problem of communications, in later sections.

A STUDY OF THE WORKS COUNCIL STRUCTURE

The second part of the communications study, to do with the structure of the Council, was taken up concurrently with the work on group relationships described above. Examination of the structure of the Council suggested that its make-up was not as suitable as it might be, particularly from the point of view of its growing aspiration towards being a policy-making body for the London factory. In composition it was essentially a Works Division and not a factory-wide Council (in the sense that its elected members came from the Works Committee); it indirectly represented Grade III Staff (who were also represented by top management) but had no elected representation at all either of Grade I or Grade II Staff (whose needs were catered for directly by top management). Moreover, while some sections of the factory did not have direct elected representation on the Council, there were other sections which were represented by more than one means: the Grade III Staff was indirectly represented both through the Works Committee and the Divisional Managers Meeting; and the Superintendents were catered for not only by their own elected representative but by the Works Manager, the Works Director, and the Chief Production Engineer as well.

Because it was not a representative body for the factory as a whole, the Council was not effectively placed to come to conclusions regarding such matters as, for example, principles covering relative levels of status and payment among the three grades of staff and the hourly paid workers, or any other matters of principle which affected all personnel. It was the Divisional Managers Meeting, and not the Works Council, which was the most centrally placed consulting body for the factory, having, as it did, independent relations with Grades I, II and III Staff Committees and with representatives of the Works Committee, as shown in Diagram Eleven.

Such a set-up had not been free from trouble. The Grade III Staff Committee, two years before, had sought independent, direct representation on the Council, but had had to accept indirect representation through the Works Committee. The Superintendents Committee had never felt easy about the role of its representatives, all of whom had experienced acute discomfort at Council meetings where there was little that could be done other than sit quietly and listen to the views expressed by the higher management members, with whom they did not feel they could properly disagree. The Grade II Staff Committee, like their Grade III confrères, also desired independent elected representation; a desire which was strengthened when the Personnel Manager confessed to them that he was himself

136

perplexed by the existing arrangements. He was supposed to be the intermediary between the Staff Committees and the Divisional Managers Meeting; was he, however, supposed to take the initiative in seeing the Grade II Staff members before each Council meeting, or was he to wait for them to get in touch

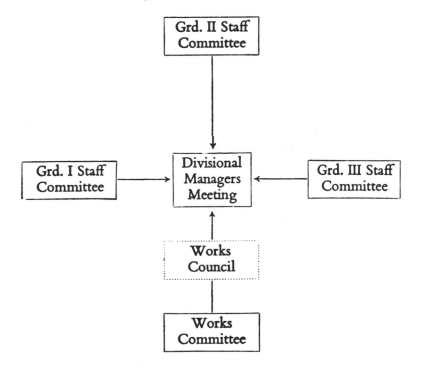

Diagram 11: The central position of the Divisional Managers Meeting in the social structure of the factory.

with him when they had any specific problems? This had never been clear.

A number of these matters came to a head simultaneously in November, 1949. The Superintendents Committee and the Grade II Staff Committee had exchanged views, and decided that their needs would be far better met if the Grade II Staff Committee were to have the place on Works Council occupied by the Superintendents' representative. A request from the Grade II Staff Committee for a place on the Works Council came before the same Council meeting—at the end of November

—as a report from the Internal Development Committee, recommending that serious consideration be given to the possibility of changing the structure of the Council in order to lay a foundation for improving communications. The Internal Development Committee's recommendation was accepted, and the Committee was instructed to go into the problem and to make specific proposals for change.

A PROPOSED NEW WORKS COUNCIL STRUCTURE

During the next three months, members of the Internal Development Committee, assisted by a Research Team consultant, met with the three staff committees, the Divisional Managers Meeting, the Superintendents Committee, the Works Committee, and the district trade union officers. They elicited the following views about the structure of the Works Council in addition to those discussed above.

The Divisional Managers Meeting thought that the management should be represented *ex officio* by the General Manager[1] alone. The other appointed management members should be replaced by direct elected representatives of the three staff committees, the divisional managers themselves being catered for by their Grade I Staff Committee for purposes of consultation. The General Manager could arrange to have any necessary managerial personnel co-opted for special purposes.

The Grade I Staff Committee was satisfied with the existing arrangements, but was prepared to co-operate in bringing about any changes if there were groups in the firm who were not content.

The Works Committee was not strongly in favour of, nor yet opposed to, changes in the Works Council, but did consider that Grade II Staff should be allowed direct representation. If changes were to be made, however, the Committee members wanted assurance that any new Council would be able to get its business done as efficiently as the existing one, without hold-ups due to frequent reference back to constituents. They also were insistent that the workers' representatives should continue to make up at least one-half of the Works Council, and that if Grade III Staff Committee were to have independent direct representation, then

---

[1] A sorting out of the role of the Managing Director had occurred at this time in the Divisional Managers Meeting, so that his role of Managing Director was distinguished from his role of General Manager of the London factories. This change will be described more fully in Chapter Eight.

Grade III should no longer vote in the shop and Works Committee elections.

The district trade union officials recognized that a Works Council on which might sit elected representatives of non-unionized staff personnel (as in the case of Grade II Staff) would be a considerable departure from the standard practice of elected trade union representatives consulting directly in a two-sided relationship with top management. But they were sufficiently in tune with developments in the firm to allow that if the workers in the factory wished such a set-up, then the trade union movement should co-operate. Perhaps a new consultative system could be developed which would be an improvement on the old.

All of the groups were in favour of maintaining the unanimity voting procedure on all matters affecting policy; and of the four groups—top management, superintendents, Grade II and Grade III—which expressed an opinion on this subject, all were in favour of the Works Council becoming a policy-making body for the London factory.

Outline of the New Structure

Putting these findings together, the Internal Development Committee and the Research Team drew up a model of the new Works Council which they considered would both meet the requirements of the various groups and provide a structure which would contribute to more satisfactory and efficient joint consultation. The proposed structure was based on the principle of each main layer in the factory's organizational hierarchy having representation on the Council, so that everyone would be directly represented by an elected representative. To carry out this principle, a fourteen-member Works Council was proposed with, very roughly, one representative for every hundred personnel, made up as follows:

Grade I Staff	(45 members) . .	1
Grade II Staff	(85 members) . .	2
Grade III Staff	(285 members) . .	3
Hourly paid operatives	(840 members) . .	7
General Manager (ex officio, representing the whole factory)		1
Total . . .		14

139

The pattern of the proposed structure in comparison with that existing at the time is shown in Diagram Twelve.

The Internal Development Committee set out certain broad principles for the constitution of such a multi-sided Council, holding that it should be a factory policy-making body, and recommending that the unanimity rule be maintained so as to eliminate questions of power politics and to ensure that there should be a full work-through of all issues in arriving at agree-

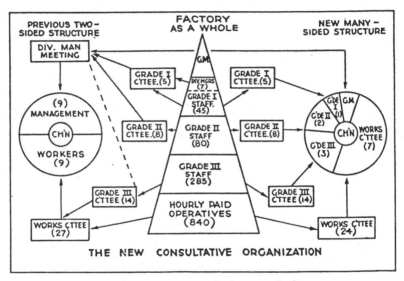

Diagram 12. New consultative organization.

ment for any particular policy. The existing method of choosing a chairman should be maintained, as should the steering committee and standing committee structures, although in view of the smaller size of the Council the five standing committees should be reduced to three in number.

The Internal Development Committee placed its proposals for a new Council structure before the Works Council on the 21st March, 1950. The proposals were adopted in principle, but their implementation was held in abeyance in order to integrate them with certain changes which had been under discussion concerning the trade union organization in the factory and the composition of the Works Committee. In order to complete the description of the alteration in the Works Council structure, it will be necessary to turn to these developments in the trade union sphere, which had begun nearly one year before.

TRADE UNION DEVELOPMENTS

In a Works Council meeting on 24th April, 1949, the management members had stated their view that the relations between the company and the trade union officials were not all that they could be. Was it not true that they could progress no faster than their relations with such important organizations would allow and, therefore, should they not at least arrange for discussions between the Works Council, the shop stewards, and the district trade union officers to see if their relations could be improved? This idea for a meeting was wholeheartedly endorsed by the other Council members, and accordingly arrangements were made, the Research Team consultant being asked to act as a kind of chairman, or mediator. The first meeting was held on 18th May, and because the discussion was not completed at that time, a further meeting was held on 15th June.

The meetings were taken up with the examination of a number of obstacles which stood in the way of co-operation between the trade unions and the management. One of the obstacles was the fact that the Works Committee, while made up only of trade union members, was, nevertheless, not the more usual type of joint shop stewards committee found in factories with a more extensive trade union membership and a higher degree of union organization. At Glacier a man did not have to hold a union card to take part in the election of Works Committee members, (with the recognized type of joint shop stewards committee only union members could take part in elections, although the committee would act on behalf of all workers, whether union members or not).

There was a strong sense of divided loyalty among many of the shop stewards. They believed that the Glacier development represented a sincere attempt to move in a progressive direction, and yet it was not as closely linked to trade union developments as they would have liked. For while the management was trying to develop and implement a progressive social policy, contact between the management and the union officials had usually only taken place when there was trouble to be sorted out. And although the management desired good relations with the unions, some of the union members had reservations about just how far it was possible, in a privately owned company, to achieve extensive management-worker co-operation through joint consultation.

Because of their lack of close contact with developments in Glacier, the district union officials admitted that they felt uneasy and suspicious about what was taking place. How could they feel otherwise, they asked, in view of what had happened in October, 1948, just before the ballot in which the vote had gone against any change to a shop stewards set-up. One of the members of the management had spoken over the loudspeaker system and had appeared to favour the anti-union cause when he said that if there was a change to a union shop, those employees who were already employed by the firm would not necessarily have to join a union in order to be able to retain their jobs. If the management was now going to help to ensure a changeover to a joint shop stewards committee, then the union officers would have reason to feel less anxious.

The management members argued in turn that the initiative and responsibility for ensuring the healthy growth of unionism in the factory must of necessity remain in the hands of the trade unions, and that the management, as such, could have no part in activities of this nature, other than to provide facilities for these issues to be thrashed out. Everyone must agree that the task of union organization at Glacier was made more difficult because the existing consultative set-up led some people to say that there was nothing to have unions for; but for the management to interfere in an issue which was for the workers themselves to settle in their own way would surely only make matters worse.

The management argument was accepted as having merit. The workers and their representatives must settle their own problems, but would have all facilities extended to them. The initiative was passed to the trade union representatives (district officers and shop stewards) to frame proposals for general principles to govern the relations between the management and the unions, and as a first step to discuss these proposals with the management with a view to getting a tentatively agreed draft document to place before the Works Council.

An Unexpected Development
Following the first submission from the trade union, a small

committee composed of the district union officers and three members of the top management (the General Manager, Works Director, and Personnel Director) was set up, and discussions went on spasmodically between October, 1949 and March, 1950. During this time, towards the end of November, 1949, an event occurred which produced a marked step forward toward the improvement of management-union relationships. The firm sent a delegation to a conference with eight other firms to discuss general problems of joint consultation and industrial relations. Each delegation to this conference had to include management, supervisory, and worker personnel; Glacier sent a party of eight, selected by its Works Council, which included the General Manager, a Superintendent, four Works Committee members, and also two of the key district trade union officials. In a setting of a week-end of intense and fruitful discussion, the Glacier delegation was stimulated to thrash out some of the trade union problems affecting the firm, and drew up the following document:

The Glacier delegation, which includes two trade union officials, unanimously agreed upon the following points and recommend them to No. 1 Factory Works Council:

1. *That the Factory Management should agree amongst themselves and make amply clear to other members of the factory, their belief that members of the factory should join their appropriate trade unions and become active members.*
2. *That because the structural details of organization of our committees, through which our union members express themselves, is of great importance (not only to our workers, but also to management itself) management should lend active help in assisting our union membership to build that organization. By this means, the democratic structure of representation can be integrated with the executive structure to the benefit of all of us.*
3. *That management should take no other active part in stimulating union recruitment on the grounds that direct persuasion or any form of coercion by management is open to serious misinterpretation.*
4. *That our trade unions should be asked seriously to consider a change of approach, from a defensive standpoint towards one of an active co-operation, by seeking for their union membership the right to a real share in the management of the factory in the form*

> *of taking an equal part with management in the formulation of*
> *such policy and legislation as members feel themselves able to*
> *undertake, leaving the executive structure of managers to manage*
> *the factory within the terms of such agreed policy.*

This document was submitted to the Works Council and its substance was adopted in the form of the following resolution, which was unanimously passed and became factory policy in January, 1950:

1. *That effective collaboration between worker representatives and factory management would make possible an integration with the management hierarchy of the current desire on the part of some for a representative structure more in line with conventional union practices;*
2. *That the roles of trade unions and their members in the future should be to seek and take full participation in management by sharing equally with managers responsibility for the formulation of such factory policy as they feel able to undertake; and*
3. *That it is the responsibility of the executive structure of managers to manage the factory within the terms of such accepted policy.*

Proposals for a Trade Union Structure

By April, 1950, just a few weeks after the Works Council had adopted in principle the Internal Development Committee's proposals for a new Works Council structure, the special committee made up of the district trade union officials and the three members of top management had completed its deliberations and had drawn up proposals governing the place of the trade unions in the firm's consultative organization. Into this plan had been incorporated all the proposals of the Internal Development Committee, so as to make one comprehensive scheme. The Works Council, to settle once and for all the various problems of organization with which it was faced, arranged an all-embracing meeting comprising the five district organizers, all the shop stewards, the Works Committee, top management, and representatives of Grades I, II, and III Staff Committees, a total of about sixty people. The meeting took place on 26th April, 1950, beginning at two o'clock in the afternoon and being scheduled to run on until agreement of some kind had been reached.

The proposals incorporated in the document before the meeting were set out in two main sections. The first section related to the structure of workers' representation in the consultative machinery, it being proposed that a shop stewards organization should replace the existing structure. All members of shop committees and of the Works Committee should be shop stewards, elected only by their respective trade union members in their departments, but acting on behalf of all workers, whether trade union members or not. It was not, however, proposed to make it mandatory for members of the firm to belong to a trade union. Secondly, the document set out a constitution for a new multi-sided, fourteen-member Works Council based on the recommendations of the Internal Development Committee, on which the Works Committee delegation of seven would include one shop steward from each of the five main unions.

On these two main issues, the decision they had to take that afternoon was whether or not they would recommend to the Works Council that a ballot be held of all hourly-paid workers to determine how the hourly-paid workers wished to be represented on consultative bodies—by means of the existing set-up, or by means of a shop stewards organization. The second, that of the Works Council constitution, was a matter affecting the whole factory, and would necessarily be dealt with by the Works Council, but had been set out to show the full implications of a shop steward organization.

The Problem of Disfranchisement

To begin the meeting, the most senior of the trade union officers reviewed the negotiations between the unions and management over the previous few years. There was steadily increasing mutual understanding, as demonstrated in the collaborative drawing up of the document which was before the meeting, and which had been arrived at not merely in a spirit of compromise, but with a real attempt to obtain integration of the views of the management and of the unions. He made a particular point of the flexibility in the outlook of all parties, as for example, in the trade unionists' willingness to co-operate in a multi-sided Works Council, even though such a set-up was far removed from normal trade union practice. They were willing to take the risk of temporary criticism from within the trade union movement if by co-operating there was a chance that new methods might be created, which would best meet the needs of the company and the unions, and which might also make some contribution to the improvement of industrial relations in British industry as a whole.

This spirit of integration, demonstrated frequently during the meeting, was accompanied by the attitude that the difficult problems still to be encountered would only be resolved satisfactorily if there were sufficient confidence among them to make for compliance and adaptability, rather than the rigidity and legalism which arose out of suspicion and mistrust.

The problem which was regarded as the most critical and about which there was least agreement, was the possible disfranchisement of non-trade union members, should there be a vote in favour of a shop stewards set-up. Some were worried because they held that any disfranchisement whatsoever of non-trade union members would be undemocratic; others equally strongly held that such disfranchisement was not undemocratic, and that, indeed, the trade union movement had been able to better the conditions of workers specifically because it had worked on the principle that when workers had an organized majority they should not allow themselves to be dictated to by a minority which did not care to organize itself and to show solidarity with fellow workers. These were widely divergent attitudes on a matter of principle. Resolution of these differences was not the primary purpose of the meeting—that was for the ballot to decide. But it was well that people should be able to express their views to each other.

Whatever the differences of opinion on the subject of disfranchisement, the trade union officials were at pains to make it clear that the workers' organization would be controlled from inside the factory. They wished it to be perfectly understood both by management and the workers that they were neither able nor willing to give day-to-day attention to Glacier's affairs, and that the responsibility and initiative for participation in factory government would have to rest with the shop stewards. The district officers would be glad to give what assistance they could, drawing on trade union resources of knowledge and experience when there were problems which could not be resolved by the people inside the factory.

Grade III Staff

Another quite different set of problems was raised by the

exclusion of Grade III Staff from the ballot and from the proposed shop stewards committee. The difficulty arose partly out of the composition of Grade III Staff with its assortment of technicians and typists, section supervisors and office supervisors, skilled workers and clerks. Such a conglomeration of personnel, created for administrative purposes, did not readily lend itself to incorporation within a comprehensive consultative organization. The situation was made even more complicated by the presence of a small minority of active trade unionists within the larger body of unorganized Grade III Staff. Matters had been brought to a head a few days before, when the draughtsmen, along with certain other technicians, who were solidly organized in the A.E.S.D., had voted in favour of throwing in their lot with the hourly-paid workers rather than with their Grade III associates. They would elect their shop steward to the Works Committee and relinquish any rights to representation through the Grade III Staff Committee.

This withdrawal of the draughtsmen had given rise to grave concern among the remaining members of the Grade III Staff Committee, most of whom were staunch trade unionists, and caused them to reconsider their position. They had been in favour of the proposed new Works Council structure because it gave them independent representation. But, if the Works Committee became a shop stewards committee, including even the representative of the only fully unionized section of Grade III Staff, they realized that the very independence which they had up till now so highly valued would also land them outside the trade union section of the consultative structure. It would be far better, they now thought, to forego the advantages of direct representation in favour of the greater advantages of the existing arrangement of being represented on the Works Council through the Works Committee. For although they had had some misgivings when they accepted this arrangement two years before, they now thought that great mutual benefit had resulted: the workers and supervisors had been brought closer together; Grade III Staff and the hourly-paid workers had been able to give each other valuable mutual support; and there was a greater spirit of harmony all round.

The trade union officials and shop stewards retired to consider this proposition from the Grade III Staff Committee. On their return, their spokesman reported that they were flatly opposed to the proposition.   They were willing to stretch trade union practice to the limit, but they were not prepared to go to the extremes of bringing non-organized groups right inside the trade union set-up.   After all, this was simply asking them to offer trade union support and guidance to people who were out of sympathy with the unions.   They did not intend to infer that the Grade III Staff representatives themselves were opposed to trade unionism, but certainly the majority of their constituents were.   Surely the true answer to the problem lay in maintaining good relations between the unions and the management and the Grade III Staff.   If relations were not sufficiently good, it was unlikely that any procedural arrangement, such as having Grade III Staff members on the Works Committee, could solve anything.   In the face of such a firm stand by the trade union officials, the Grade III Staff Committee members agreed to drop their request, and to report back to their committee the views which had been expressed.

Deciding on a Majority

The meeting then returned to the main question—whether it should recommend that a ballot be held.   A motion was proposed, and unanimously endorsed, that the Works Council be asked forthwith to consider holding a ballot to determine whether the hourly-paid workers wished to change their form of representation to a shop stewards set-up.   Further, before such a ballot was organized there should be full discussion of the proposals both with members of the factory who held trade union cards and those who did not.

But what size of majority was to be considered necessary to bring about a change?   They were dealing with an issue of personal and vital importance to each hourly-paid employee. Hence, they all readily agreed, it was essential that well over 50 per cent. of the people concerned should be in favour before any change was implemented.   The question was, how much over 50 per cent.?   Some argued for 75, others for 66 per cent. in favour.   No one knew of any consistent precedent from outside the firm which could help them settle the issue, so they decided to adopt the size of majority used in the previous ballot on the same issue in October, 1948.   To their surprise, this turned out to be a lower figure than anyone had put forward—namely, 60 per cent. in favour—and this was the figure settled upon.

The meeting then broke up just after 8 o'clock, slightly over six hours after it had begun.

THE NEW STRUCTURE IS ADOPTED

The report from the trade union meeting placed before the Works Council on 4th July, contained support for the setting up of a multi-sided Council of fourteen members which would be a part of the policy-making network for the company. Of the fourteen members it was proposed that seven should be from the Works Committee with at least one member from each of the five unions—the A.E.U., A.E.S.D., A.U.F.W., T. & G.W.U., and E.T.U.

The document on which the proposals were set out had been printed and 400 copies distributed throughout the works, nearly one copy for every three people. The Works Committee, the group mainly affected, had not organized any very extensive discussion in the shops. They reported that there was no high interest, and the workers had not spontaneously sought to get copies of the document, although these were available. There was said, however, to be a lot of animation which was not being officially expressed.

In contrast to the hourly-paid workers, the staff committees had shown keen interest. The Grade I Staff Committee, concerned about what was meant by the Council becoming a policy-making body, called a general meeting of all its constituents, and drew up a number of amendments; the Grade II Staff Committee discussed and accepted the document; the Grade III Staff Committee was taking steps to discuss the matter with all of their constituents and had not quite completed the task.

Two of the amendments put forward by the Grade I Staff created considerable discussion in the Council. The first of these had to do with the size of quorum for Works Council meetings. As proposed, the quorum was to be composed of nine members, of whom four should be members of the Works Committee, but the Grade I Staff wanted this extended to include at least one representative from each constituent body. The Chairman indicated that such a requirement would prevent the Works Council from meeting if the one representative of Grade I or the General Manager were absent. This was exactly the point, explained those who were speaking on behalf of the Grade I amendment, for if the Works Council was to be a policy-making body, then they would not like to see the possibility that policy could be decided in the absence both of the General Manager and the Grade I Staff member. The issue was finally resolved by modifying the amendment to a requirement that either the General Manager or the Grade I Staff representative be present for a proper quorum to be constituted.

The second difficult point had to do with the unanimity voting rule. Grade I proposed that the existing position requiring unanimity only on matter of policy and not on matters of procedures should be changed to that of unanimity being required except on a question of procedure. This change would have the effect of ensuring that where there was any doubt whether a motion was on procedure or policy, it would be treated as policy and a unanimous vote would be required. The heated discussion on this point lasted well over an hour. So many other problems got mixed in that one Councillor was constrained to observe that perhaps it would be better if the Works Council stayed as it was and they just forgot about revising the constitution. The new proposals, he thought, had raised far too much suspicion. He was correct in the sense that there had been revealed a great deal of suspicion between members of the Council, and between the different groups they represented in the factory.

The two Grade I Staff Committee amendments were intended as checks to prevent the Council from taking precipitate action against the best interests of the factory, and this was perceived by other members of the Council as a lack of confidence in the representatives of other sections of the factory. What was not recognized, however, was that the Grade I Staff Committee had by-passed their normal route to the Works Council, which was via the Divisional Managers Meeting. It may also be noted that the Divisional Managers Meeting had worked out their own views about the new Works Council structure with little reference to the views of their immediate Grade I subordinates. Because the members of the top grade of management found it difficult to communicate with each other as executives about joint consultation, they took the step of using their Grade I Staff Committee to protect their executive authority. It could be said that the Grade I members feared that the General Manager might allow too much executive authority to slip through his fingers, but that they were unable to express such fears directly through the executive chain.

The Grade I amendment on unanimity produced much consternation. A series of amendments to it were defeated one after the other without any satisfactory formula being discovered to resolve the points of disagreement. There was nothing left for the Chairman to do but to put the original amendment to the meeting. It was a matter of policy, he said, and must therefore have a unanimous vote. They should also keep in mind, he warned, that the Grade I Staff Committee members could not vote in favour of the document if the amendment was defeated. This would mean that the whole document, with the months of hard work that had gone into it, would have to be scrapped, and they would have to start all over again. He then put the amendment to the vote and, in spite of his warning, it was defeated.

In order to overcome the deadlock and the delay which loomed ahead, the General Manager asked if they could not even at this late stage hear why members had voted against the amendment. It then turned out that the Grade III members, who had voted against the amendment, were not opposed to it at all. They were taking this action as a last resort to prevent their exclusion from the Works Committee, which would happen if the Works Committee became a Shop Stewards Committee. Just prior to the Works Council meeting, the Grade III Staff Committee had met and reversed its previous decision to seek independent representation on the Council. But this, the Chairman admonished, was surely for the Grade III Staff to thrash out with the Works Committee and the trade union officials, and was not the kind of issue on which they could fairly hold up the Works Council business, as they were doing. Satisfied that they had been heard, the Grade III Staff members concurred, and the amendment was again put to the meeting by the Chairman, and passed.

The impression of Research Team members following this meeting was that the Works Council had lost sight of the major objectives which it was setting itself in establishing a new multi-sided structure. The reason for this seemed partly at least connected with one of the Council's most immediate communications problems—that of the capacity of its members to speak their minds freely to each other.

Grade III Staff and the Trade Unions

The Grade III Staff Committee straightened out its difficulties at a special meeting of the Works Committee on 14th August, attended by the district trade union officials. The Grade III members were strongly in favour of giving up their independent representation on the Council and maintaining their existing relationship with the Works Committee. They had no new reasons to put forward other than those which they had stated the month before at the meeting between the Works Council and the trade unionists. There were many members of Grade III Staff who were staunch trade unionists and wished to maintain their trade union ties; they felt anxious lest under the new set-up they should be pushed out. Had not relations between Grade III and the hourly-paid operatives been improved by the working together made possible by having Grade III representatives in the Works Committee? The supervisors did not want to be placed on the management side; they wanted to be a bridge between the management and the workers.

The difficulties, however, were considerable. If the Works Committee became a shop stewards committee, the Grade III Staff Committee members could only be elected by trade union members so as to hold shop stewards' cards; but this would mean the Grade III Staff Committee being elected by a small minority of Grade III members. No easy way out could be discovered, so they decided the best thing to do was to go ahead and hold the ballot on the shop stewards question, and to set up the new Works Council. If the ballot went against the shop stewards set-up, then the existing arrangements would be maintained; if, however, the Works Committee did become a shop stewards committee, then the Grade III Staff Committee should take its three independent seats on the new Works Council. They could then open negotiations with the Works Committee for the purpose of arranging for co-operation between the two bodies.

The New Works Council and its Functions

At the meeting on 22nd August the new Council constitution was adopted, with a few minor amendments, and passed on to a drafting committee for final editing. On 20th September the ballot on the structure of the Works Committee was held. The result was 61.5 per cent. of the votes cast in favour of a shop stewards set-up, and 38.5 per cent. opposed, so that by a narrow margin the shop stewards set-up was adopted. In all 89 per cent. of the hourly-paid operatives cast their votes. Following the ballot, the Works Council set 28th November as the date on which the new Council would come into being, the intervening two months being set aside to allow the Works Committee to carry out the elections for the new Shop Committees and the Works Committee, and so that the latter body could choose its seven members for the new Council.

During the course of the above described discussions beginning with the week-end conference the year before, the functions of the Council had become more clearly specified and formulated. The most precise statement of Council functions was made in the trade union document and, as modified by a special report from the Internal Development Committee, appears in the new Works Council constitution in the following form: "The Functions of the Council shall be: . . . to carry the responsibility of deciding the principles and policies which shall govern the Management of the Factory in the light of opinions of producers and managers, in the light of the interest of consumers, shareholders and the nation at large, and in the light of total Company Policy. To revise and clarify, where necessary, old policy in the light of the opinions and interests mentioned above. . . ."

As though to give substance to this formula, the Council at its August meeting tackled two major questions of policy which had been thrown up at the Works Conference in May. One, concerned with redundancy, was thoroughly aired, and gave the management an opportunity to state its own recent thinking on the matter. The outcome was the sending of the following resolution to the Board of Directors:

"That redundancy should be minimized, and towards this end there should be an examination of the possibilities of:
1. Redisposition of the Labour Force.
2. Distribution of the Company's products.
3. The organization, on a very long term basis, of additional Company factories making bearings and allied products, with the object, amongst others, of creating the situation where the load could be interchanged between units of the company.
4. Stocking up of Service Department."

This resolution was subsequently considered and adopted by the Board, and became part of the company policy. The second issue had to do with the wages policy of the factory as a whole, with particular reference to the principles governing the relative wage levels of the various grades and groupings of personnel from top management to shop and office employees. To the Conditions of Employment Standing Committee was delegated the responsibility of seeing that a proper investigation of the problem was carried out. A number of other matters of policy of this magnitude remain on the agenda of the Council, and the new Works Council will take over a burden of exceedingly difficult work.

CONCLUSION

The Works Council study began with the problem of how to improve the business procedures of their meetings, and from this beginning there emerged a system of standing committees which would help the Council to conduct its business with the greatest possible efficiency. Consideration of the functions of these standing committees was made difficult because the question of the authority and power of the Council became mixed in. It was not until the difficulties over these more basic issues of power and authority were tackled openly that the rather more simple problem of setting up standing committees could be settled. Similarly, difficulties over the time and cost to individuals of participation in joint consultation turned out to be only partly a problem in its own right, and partly a symptom of the Council members' lack of security about their prestige and status. Once this symptomatic aspect was recognized, the cost problem was more or less easily dealt with. The recognition of these anxieties about status and prestige, and of these problems of the power and authority of the Works Council, led to the holding of a week-end conference, at which some of the basic principles of the Council's work and purpose were faced and clarified.

One of the results of the week-end conference was the initiation of a study of communications. This study gave indications that barriers to communication came about not so much through failure of individuals to carry out their job as through fears about the too free communication of feelings which were sensed as being destructive. Because of these fears, the executive system was kept tightly closed to upward communication. The Works Council members by unconsciously keeping themselves walled off from the rest of the factory were able to protect themselves from too much outside disturbance and maintain the Council as a kind of segregated consultative club.

As part of the communications study, the structure of the Works Council was examined, to determine whether it was as well constructed as it might be from the point of view of establishing relations with constituents. After an extensive work-through lasting nearly a year, the Council was changed from the two-sided Works Divisional Council which it had really been, to a multi-sided council composed of representatives of all strata of the factory. Concomitantly with this shift in structure, the purpose and functions of the Council were settled, it being made part of the policy-making network of the company as a whole. By the difficult process of working-through to an all-factory representational body, it had become possible to create a sound basis for the democratic sanctioning of executive authority.

CHAPTER SIX

THE WORKS COMMITTEE

Relations Among Workers' Representatives

IN times of stress and crisis, when a group of workers and its leaders are experiencing a common sense of danger or insecurity, it is not difficult for the leaders accurately to reflect and express the group point of view. In contrast, times of relative calm and stability pose greater difficulties for elected leaders and make greater demands on their representational skills. How much this latter difficulty is likely to be reduced or increased where there is the possibility of genuine co-operative attitudes within a factory remains a question of importance. The general situation at Glacier gave the opportunity to consider this question because of the opportunity offered to the workers to have a real stake in setting the general policies governing the factory.

The material selected for consideration deals mainly with certain variations in the quality and character of the work done by the Works Committee as the general conditions in the factory varied. A few specially chosen events will be taken in order to throw some light on why these workers' leaders encountered such profoundly difficult problems in carrying out their tasks in a setting of constantly improving worker-management relations.

The leading members of the Committee had had from ten to thirty years of experience in the trade union movement, and included representatives whose working-class background was that of Clydeside, Tyneside, and North Wales. According to the judgment of those members of the Research Team with wide experience of the trade union movement in England and Scotland, this was a group of leaders of considerable ability for a medium-sized factory, and their difficulties could be said to derive largely from the peculiar circumstances in which they found themselves rather than from lack of ability or militancy.

GENERAL IMPRESSIONS OF THE WORKS COMMITTEE

In January 1949, the Research Team received an open invitation for its members to attend the Works Committee meetings. At that time the Committee was composed of twenty-four representatives of hourly-paid operatives, plus three representatives elected from the Grade III Staff Committee who had just been added to the Committee the previous November in order to give Grade III Staff a consultative route to the Works Council. All committee members had to be bona fide members of a trade union, and about half the Committee held shop stewards' cards. The relations between the Committee and the shop steward group in the factory had been left unsettled by the November 1948 ballot when a proposal to change to a shop stewards set-up had been voted down.[1]

The objects of the Committee, as stated in its constitution, were "to provide a direct channel of communication between employees and the management; to provide machinery for the joint consideration of such matters affecting the employees and the factory as may be suitable for joint discussion; to provide the means for constructive co-operation in obtaining efficiency, and the comfort and well-being of those employed by the company; and to deal with all reasonable subjects affecting the members of the factory (with the exception of wage rates, which are controlled by national agreements)".

The Research Team attended regularly as visitors from the time the invitation was received—being the only outsiders present since the meetings were held *in camera*. Attendance, as observers, with minor contributions from time to time to the business of the meeting, went on until nearly a year later, when a more active collaborative role was undertaken in connection with the project on communications which had commenced in the Works Council.

Attendance at the Works Committee produced two definite but completely opposite impressions in the minds of the Research

---

[1] This ballot, described on p. 44, is not to be confused with the more recent ballot conducted in September 1950, described on p. 152 in which by a narrow margin there was a vote in favour of a change over to a shop stewards set-up.

Team. From January until April, the period of the redundancy crisis, one listened to serious and earnest discussions about redundancy policy, and gained the impression of a committee engaged in tackling problems of first importance to the workers in the factory and safeguarding their interests. They talked over how to get jobs elsewhere for workers who left, and job security for those who remained; and when the redundancy crisis deepened by March to the point where it was thought that really valuable members of the firm were being lost, there was a review of the principles of choosing people for redundancy, with particular reference to the wisdom of continuing a policy that did not conform completely with the standard trade union practice that married women should be made redundant before men.

In April the redundancy crisis diminished in intensity and the employment situation stabilized. In September, following an upsurge of orders spurred on by the devaluation of the pound, more workers had once again to be taken on. Along with this, a change took place in the character of the Works Committee meetings. The Research Team members were now left with the impression that the matters dealt with in the Committee had little to do either with the discussions they had heard taking place in the Works Council or with the matters which they heard talked about from time to time in various shop committees and at shop floor level. As a factor in communications between the shop floor and the Works Council, the Works Committee was acting more as a barrier than as a channel.

To take some examples, during the summer and autumn of 1949 the Committee spent a great deal of time discussing whether they should agree to the works taking part in a blood donors' scheme, and whether the workers should subscribe to a children's charity. While they were giving priority to such items—which, whatever their intrinsic merit, were of only secondary importance to the workers—there were many other events of first importance taking place which the Works Committee never talked over; for instance, the change-over in methods of payment in both the Service Department and the Foundry. Such failure to discuss payment questions which would undoubtedly have extensive and long-term effects, was particularly striking in view of the continuous complaints of the Works Committee members about the apathy of the workers they represented. In the same way, although they frequently criticized the wide range in the quality of supervision in the factory and variations in the degree of co-operativeness of supervision with shop committees, they did not seriously get down to talking about what they might do about it. That they were hard put to it to deal with the things that the operatives, including themselves, were chiefly concerned about was, however, either not recognized or else ignored.

As another illustration of this detachment from critical shop floor issues, there was a noticeable lack of reporting about what was going on in various departments, it being only on rare occasions that the members gave any indication of salient happenings in their shops. Similarly, there were hardly any reports on issues being taken up by the Works Council, or any prior discussions of the agendas of impending Works Council meetings. In the absence of reports both from the shop floor and from the Works Council, and without much discussion of Works Council agendas, it can hardly be said that either the Works Committee or its representatives on the Council were acting as representatives of the shop floor; each individual really knew very little about any other shop than his own.

These differences in the behaviour of the Works Committee in times of crisis and times of relative stability were striking: in crises, the Committee mirrored the fears and wishes of its constituents with great clarity, and hard work and action resulted; in periods of calm, the Committee to a greater or less degree lost touch with its constituents, and became a more or less circumscribed group of individuals—a kind of consultative club— with few apparent influences on it from outside.

THE TIME AND COST OF JOINT CONSULTATION

Accompanying these difficulties of the workers' representatives in carrying out their functions, was the long existing problem of the cost both in time and money to individuals who took part in consultative activities, those who were on piece-rates losing anywhere from 2s. to £1 per week in bonus earnings. This problem had become particularly acute between January and April 1949, during the redundancy crisis, when the Works Committee representatives had been heavily committed in reviewing all cases that had come up as redundant. They had been off their jobs for many hours each week to protect the interests of their fellow workers by making sure that the principles for carrying out the redundancy procedures were thoroughly adhered to; and this responsibility, because of the painful nature of the problem, had carried with it a considerable emotional strain. Added to all this there was the burden of meetings— shop committee, Works Committee, Works Council, trade union branch, and others—held in the evening after working hours, and interfering with home life and leisure time.

In June and July 1949 three Works Committee members resigned. The reasons given were the excessive amount of time required, with consequent loss of money; the danger of losing one's skill through being away from the job; and the fact that the supervisors tended to put them on less important jobs because of the uncertainty of their attendance, and this in turn further contributed to loss of earning and skill. Two of those who resigned had long records of active service with the Committee and were on the Works Council, so that their resignations caused much uneasiness.

On 19th July the Works Committee was talking over what it should do about this problem of resignations when the Chairman asked the Research Team consultant if he had any comments to make, since the Research Team were engaged on a study of the time and cost of joint consultation. The consultant replied that he was willing to try to shed some light on their difficulty but warned that he would have to range over a wide territory and might bring in some unexpected factors, and even then might not be able to provide them with any easy solution. In the first place, he said, he noticed that they usually discussed their problem in terms only of the cost in money, whereas it had for a long while been the impression of the Research Team that the financial problem was only a minor aspect cloaking other more important factors. For, if the money side had been a prime consideration, it would be difficult to explain why nothing was ever done about it; there should have been little trouble in ensuring that nobody who took part in consultative activities would be out of pocket. And in the same connection how was it, he asked, that many of the Works Committee members did not even take the money to which they were entitled in A.W.I.s under the then existing arrangements? Behind the inability to do anything about. the money question there was a feeling of hopelessness about the value, except in times of crisis, of their

160

consultative activities, and if this were true, as he thought it was, then it did seem as though, in order to escape from despair, they preferred to go on saying that their difficulties were due to the money loss. In short, everything was blamed on money, but nothing was done about clearing up the money problem because that would have meant examining the underlying causes of their despair. If such a notion was surprising to them, they had only to think back on the number of times they had themselves complained that they did not gain any prestige among their fellows as a result of the hard work they did, nor had they even received much appreciation for the long and difficult hours they put in during the redundancy crisis.

These comments were followed by a depressed silence in the meeting, and the consultant, taking this silence as an expression of their sense of futility, went on to ask whether part of the problem was not related to the fact that, since there was no real fight on with management, they felt that they were being asked to occupy a role so tame that they were rather ashamed of it. It seemed to bring great satisfaction and a sense of achievement, he added, to organize people into trade unions and to get workers' support when there was a fight on and everybody was dissatisfied with the management. But many of them had complained from time to time that they had come to be known and treated as "governor's narks", "the Gestapo", and "people who get into the supervisor's office and smoke cigarettes with the boss", and had experienced considerable anxiety about the loss in prestige and esteem which accompanied election to committees. When there was fairly good co-operation such as they had in Glacier, and they were not leading a fight, then was it not far more difficult for those in representative roles to appear as though they were really doing a job on behalf of the workers?

This in turn came right back to the very problem of payment for time spent on consultative work, which they had been considering. For if it was true that they felt their efforts were futile when there was no actual crisis to deal with then did it not look as though they clung to an arrangement for financial compensation by which they were slightly out of pocket, in order to escape any possible accusation that they were just making a good thing out of their job as representatives? If they could at least say and feel that they were suffering certain disadvantages because of their role, then they need not worry so much about whether they were doing anything of significance or not, and could hold up their heads in the shop.

The consultant did pause a number of times during this discourse, but there was little response or lively comment, even when he specifically referred to their despair and depression then and there, and related it to the absence of a clear cut purpose. At the end of the meeting, however, there was a buzz and hubbub, people broke up into cliques, and everyone went away talking actively.   This behaviour strengthened the consultant's impression that one of the difficulties of the Works Committee was the fact that it was so broken up into small cliques, each of which was out of contact with the others.   But he had not succeeded in taking this point up effectually.

A CONSIDERATION OF THE WORKS COMMITTEE PURPOSE

These extreme variations in the functioning of the Works Committee, both in the matters it took up and in its contact with the shop floor, raised the question of what functions the Committee could usefully and properly perform in between times of crisis.   This question came out sharply in the confusion of the Committee's full-time secretary about his own role.   The secretary in office had been elected when the post had first been created in 1946, but since that time had never been given clear terms of reference.   How he spent his time and what he did were left very much to him to determine.   The jobs which seemed to come his way were such things as: organizing elections in the shops; getting out notices relating to the Works Committee; getting secretarial jobs done for shop committees; acting as a link between the shop committees and management in the shops, as, for example, when a shop committee wanted to invite a member of the management to its meetings; suggesting items for the Works Committee agenda; taking up many individual cases of appeals; and acting as a link with the Works Council.   He was a kind of "all purposes man", as he put it.   He acted as an adviser, both in an official and unofficial capacity, to shop committee and Works Committee members; and he gave support to anyone who wanted his help, entering on the scene, for instance, when a shop committee or Works Committee member did not have the time or felt unable to handle a particular situation.

162

In hardly any of these activities, however, was it clear to him whether he was acting in an individual or an official capacity and, as far as could be determined, he seemed to spend more than half his time acting unofficially. He would go into a department when asked, discuss difficulties with shop committee or Works Committee representatives, and help them take up matters with shop supervision or higher management; or in an unofficial capacity, he would approach higher management or shop supervision directly on problems which he heard about via the "grape-vine", and get them sorted out "behind the scenes".

Because he was doing so many of these jobs in an unofficial capacity he had become the carrier of many of the unexpressed or unrecognized stresses between the workers and the management in different departments and in the factory at large; and there was nothing he could do but carry all of these difficulties piled up within himself with no official means for getting rid of them. If the problems were resolved, then he was not too badly off; if on the other hand the problems remained unresolved or only partially resolved he automatically remained the storehouse for the stress. Added to all this was the fact that his salary was paid by the management. This made him a focus of suspicion for some of the workers, who felt that he did more for management than he did for themselves. No wonder then that from time to time he would feel extremely uncomfortable about the position in which he was placed.

The matter of the functions and purposes of the Works Committee was brought to a head in the autumn of 1949, following the Works Council's week-end conference when the decision had been taken to change the Council into a policy-making body. This decision had thrown many of the Works Committee members into something of a panic, and had led to a compulsive demand to know more about the way the factory was run; for if the workers were to carry out their responsibilities in such a policy-making Council, they thought they would have to understand far more than they did about the various management functions. As it was put by one of the members of the Works Committee, "If the workers are really to take part, then they have to prepare themselves to become members of the Board of Directors".

The provision of opportunities for workers to study the functions of the various executive branches was the subject of one of the motions passed at the Works Council week-end conference in September 1949, it being recommended that the Personnel Director should see whether suitable arrangements could be made.   Members of the Training and Upgrading Department of the Personnel Division were put on to the job and, under the impression that the workers' leaders were most anxious to begin such training right away, brought forward to the Works Committee in November various proposals for the desired training courses.   These proposals centred around courses of lectures which could be given by the divisional managers and other managerial personnel, along with opportunities to visit the different divisions to see the work in progress.   Curiously, however, after a discussion of these schemes the whole idea was rejected.   Much to the surprise of the Training Superintendent, who came down to explain the different plans which might be arranged, the tenor of the discussions made it appear as though it was the training Department which was trying to sell a training course to the Works Committee although in actual fact it was the other way round.

Nevertheless, the subject of training in management practices for workers' leaders was not dead.   One month later the subject was raised again by one of the representatives of the Grade III Staff who had been absent at the previous discussion.   He was one of those on the Works Council who had been most anxious at the week-end meeting to see such training instituted, and he now argued that if all members of the Works Committee were not to be equally trained in all functions of management, then at least some of them should be selected to become specialists, knowing between them about the different management functions.   Specifically he proposed that the Works Committee should set up a panel of, say, seven to eight people, each of whom would be responsible for understanding the functions of one division of the factory, so that whenever the Works Committee was faced with any particularly difficult topic they would be able to refer to this panel for expert advice.

The pros and cons of this proposal were vigorously argued, and the decision not to have anything to do with management training was reversed. The new plan to set up a panel was adopted in principle, and the Committee Executive, composed of the Chairman, Vice-Chairman, and Secretary, was instructed to go into the plan in more detail and bring back specific recommendations as to how such a Panel should be formed. But this preoccupation with management training was at least partly a method of evading the Committee's difficulty in carrying out its primary task—that of representing workers. This became more or less recognized during discussions which will be described below, and the whole matter was once again dropped, It was realized that learning more about management functions does not necessarily make a person more capable of maintaining two-way contact with the people who elected him; the job of representing the attitudes of workers is one which calls for its own skills.

BLOCKAGES IN COMMUNICATIONS

In October 1949, each member of the Works Committee was interviewed in connection with the study of factory communications initiated within the Works Council. As a first step these interviews were directed towards ascertaining the views of the Works Committee on the effectiveness of two-way communications between the members of the Committee and their elected representatives on the council.

Despite the limited objective of the interviews, they tended to range over much wider issues. Nearly everyone referred at some stage to the apathy and lack of interest of the factory personnel. Many of them held that the operatives were more or less unconcerned about things unless they were directly affected. To some extent the progressive management policy was held to be partly responsible for this attitude, for "if management were to behave as they do elsewhere, people would soon sit up and take an interest in what their elected representatives were doing". Adequate facilities for the exchange of information of all kinds between all levels did already exist, but these were not used. This lack of interest was at the bottom of the communications problem, it being more a case of people not wanting to know than of not knowing.

While there could, in view of the scale of the problem, be some justification for Works Committee members not being fully in contact with their constituents, there was really no excuse, they thought, for their being more or less out of touch with the Works Council, and even less for being out of touch with the other members of the Works Committee itself. But, this being the case, how could you get any insight into what was going on in other shops in the factory when you did not get reports about these matters at Committee meetings, and hardly even knew the names of some of the other members of the Committee?

Because of their uncertainty about what exactly was the attitude of higher management and supervision towards joint consultation, members of the Committee were to some extent unwilling to try to do a good job of two-way communication. They were insecure in carrying out joint consultative activities because their impression, whether accurate or not, was that their supervisors still regarded joint consultation as an activity which had to be supported because it was company policy, but which was almost unconnected with the factory's real activity—the production process. It was not only supervisors, however, who came in for criticism, the workers themselves being regarded as not taking joint consultation seriously; time spent away from production by Works Committee representatives was perhaps even more resented by their own shop floor colleagues than by their supervisors. This resentment was particularly evident when representatives had some adverse decision to report back. Under such conditions, the Works Committee members asked, could you possibly feel that your efforts were really appreciated?

Works Committee: 19th December, 1949

In December 1949, the Research Team circulated to the Works Committee a report which pulled together the findings from the interviews carried out in October and the impressions derived from attendance at their meetings over a period of nearly a year. Although the report was very short, it led to extensive discussion throughout two meetings, one on 19th December, 1949, and the second on 11th January, 1950.

The Works Committee meeting on 19th December was the last before Christmas. There was a fairly full agenda, but as some members had to get away early, they decided to try to finish at half past seven, half an hour earlier than was customary. The report came up for discussion at just after half past six, and was taken paragraph by paragraph. A member of the Research Team was co-opted shortly after they began on the report, in order to elucidate the points raised.

"It is our impression that there is a block in communications between the Works Committee and its representatives on the Council, and that this results in the Works Committee acting as a serious block in communications between the Works Council and the Shop Floor.

(a) Little real impression is conveyed about Works Council problems to non-Council members of the Works Committee.

(b) The Works Committee itself gets practically no co-ordinated information about what is going on at shop floor level, and hence is in no position to see that the Works Council is fully briefed on shop floor feelings."

This was not true, argued some members; for was there not a considerable amount of reporting between the Works Council and the Works Committee? There was the news bulletin which gave details of Works Council discussions, and when necessary, there were reports back to the Works Committee on important items. With so much reporting, not to mention the fact that Works Council meetings were open to everybody, surely anyone who complained about lack of communication had no one to blame but himself.

The consultant emphasized that he wanted to make a clear distinction between, on the one hand, the act of reporting, and, on the other, communication which really counted—in the sense of people helping each other to get a broader perspective, so that decisions could be taken with greater understanding both of the general situation and of each other's views. It was in this wider sense that it could be said that a main source of obstruction in communication was within the Works Committee itself. For except when from time to time crises arose on special issues such as holiday arrangements, could they really maintain that they did very much in the way of exchanging views and comparing notes on matters affecting workers? Without this, their nine members on Council were in most matters reduced to acting as individuals; it being practically impossible for them to act on behalf of the Works since there was no mechanism by which they could get to know what was going on in shops other than their own.

The Tool Room representative strongly agreed with this, holding that these blocks certainly existed. But he thought that possibly some of the trouble was due to the fact that the Works

Council members of the Committee were so overburdened with duties in so many different spheres that they could not carry out their responsibilities in a way that gave them any satisfaction. The Service Department member, also agreeing, said that it was standard practice at trade union branch meetings for shop stewards to report on important developments in their own shops and to keep each other well informed about those things which were causing unrest among their fellow workers. The Works Committee was a trade union committee, and he could not see why they had not already adopted what were well-tried union methods in carrying on their business. He thought that some procedure by which members of the Committee would report on their departments was essential if they were to know sufficient about what was going on to be able to get things done on behalf of those who elected them.

"It is further our impression that the break in communications in the Works Committee stems from a breakdown in relations within the Committee itself. While other factors undoubtedly play a part, until the Works Committee gets its own affairs right, it will be difficult to straighten out the problem of communications between Works Council and the Shop Floor.

(a) The Committee seems to be split into factions, as, for example, members of one union versus members of another; those in favour of a shop stewards' set-up versus those who are not; certain groupings based on a banding together of individuals.

(b) The Committee is so large it is difficult for many members to make any contribution. In fact it seems few people even know the names of all the others, or the shops they come from."

The Chairman questioned whether there was really anything much that could be done about matters of this kind. Although such tensions as were described did undoubtedly exist, in his opinion they were largely personal matters, and had little to do with the relationships between groups. Thus, for example, some people might feel that the Research Team consultant himself was not to be trusted. They might say that he tried to smooth his own way with the management. But you could not do much about this, because such feelings had so largely to do with personal reactions, and it was not easy to change people.

The consultant accepted this example of attitudes towards himself as a very useful illustration based on actual fact. He had no doubt that there were members of the Committee who did look upon him as smoothing his way with the management. But if they examined this they would discover that there were general issues at stake as well as personal feelings. For the Research Team had come into the factory with the independent agreement of both the management and the workers. If, therefore, some members of the Committee now thought that the Research Team consultant was more on the management than on the workers' side, did this not indicate, besides personal suspicion of the consultant, suspicion that top management was the kind of group likely to try to get the consultant under its thumb and use him for their own purposes? In other words, here was an indication that the management-worker split still existed in the minds of members of the Works Committee, in spite of their protestations that they had eliminated the "two-sides" attitude in the factory, and despite their serious attempts to achieve a harmonious atmosphere based on co-operation. Most personal attitudes of this kind, the consultant concluded, could, in his experience, be looked into in the same way—for whatever could be learned from them about the stresses between groups.

One of the Committee members said that in his opinion these splits certainly did exist. But, he urged, should they not try to put them on one side and forget about them, and get on with their duties instead; after all, such splits were not worthy of workers who were supposed to be able to co-operate with each other in a united way. But another of the shop stewards from a different union said that it was not so easy just to push such matters aside and disregard them. In his estimation one of the unions, because it was so much larger than the others and had a branch in the works, tended to dominate the proceedings at Works Committee. Whether his feelings were really justified or not he would find it hard to say; but nevertheless he did feel that way, and he could not just forget it because it made him diffident about saying very much; his union was so much smaller that it made him feel out of things.

The consultant observed that further examples could easily be added to those already mentioned. There were the various cliques, for instance, which they all knew about very well and which had grown up around differences in attitude among their leading members, some being far more suspicious than others of the social developments in the factory. One of the two main cliques regarded the other as "soft", and in return was regarded by the other as prejudiced and unnecessarily destructive. Then there were the differences, which had not really been resolved, between those who felt that the workers had to accept responsibility for policy-making on the Works Council, and those others who felt that such a move was premature and was putting too great a strain on the workers. Or along another line altogether, they had the banding together of those who worked at, or represented people who worked at, so-called "dirty" jobs, such as in the Foundry, who felt that they were not given equal status, and that their needs and interests were not given due consideration and weight.

The main point, however, the consultant continued, was that such differences and feelings were an inevitable accompaniment of the activities of any group. The most useful question for them to consider, therefore, was whether or not these differences were so intense as to render it difficult for them to carry on their work in a reasonably efficient way. It was his impression, and this was why he had raised the matter in his report, that although they agreed very readily on most of their matters of business, underneath the surface there seemed to be many points of disagreement that were either not recognized or were just not being talked about—all of which was definitely impeding their work. This, he thought, had contributed to their inability to communicate easily with each other about what was going on in their shops. Or, to take a more striking example, in October 1948, when the ballot was held on whether or not there should be a shop stewards' set-up in the factory, the inter-union rivalries among the shop stewards of the various trade unions had been sufficiently strong to prevent them from co-operating to explain to the workers in the factory what was at stake, and to rally their support.

The Foundry representative quickly took up the point about reporting on shop activities. He and his colleagues from the Foundry experienced great difficulty in getting up the courage to report what was going on in their shop because, if they reported that anything good was happening, it would seem as though they were boasting. Surely there was something wrong amongst them if it was possible for some to feel that way. This point was picked up by other members who felt the same way, but the time set for the end of the meeting had come. It was only with some effort that the Chairman was able to bring the discussion to a close, because everyone had become very intent on the matters being talked about. In view of this interest, and because they had then only covered half the report, they arranged an extraordinary meeting soon after the New Year to complete their deliberations.

Extraordinary Meeting: 11th January, 1950

One of the reasons that an extraordinary meeting had to be called was that two months before, in October, the Committee had changed the times of meeting from once every fortnight to once a month. This would have meant waiting until near the end of January to complete the discussion of the report, and the members were anxious to continue.

"Because of failure to build satisfactory internal working relations, the Committee has been only partially able to discuss shop floor problems; for example:

(a) Two departments have switched their methods of payment. The implications for other workers of these changes have not been discussed at Works Committee.

(b) The Works Committee tends to blame the shop floor for being apathetic. It is said that others are only interested in their pay packet and in having a secure job. And yet methods of payment, rate fixing, and employment security are rarely taken up by the Works Committee.

This raises the question of the purpose of the Works Committee. Listening to meetings, except in times of crisis, one gets no really clear-cut impression of why the Works Committee exists. Thus, for example, there has been much discussion about training in what management does, but no discussion of training in how to represent the views of a body of workers."

The delegates from the Foundry, referring to the first example, reiterated that the change-over in method of payment in their department had gone quite satisfactorily, but they did not want to report it to the Works Committee because as they had mentioned at the previous meeting they did not want to appear to be claiming that they were running their own show any better than the others. The consultant picked this up as an indication of an attitude that seemed to be widespread in the committee; they did not have enough confidence in each other to be able to speak freely without having to think first whether what they said would be tolerated by the others.

Many members of the Committee, however, were worried that if they were to have reporting from every department they would never get any work done. These fears were answered by others who thought that it was not so much a question of detailed reports from each department at every meeting but of having sufficient reporting of important happenings to ensure that the Committee did deal with the things that mattered. For after all, could anyone honestly say that he was satisfied that they were getting sufficiently important matters referred to them from the shop floor? The Chairman commented that, in his estimation, something seemed to have happened to the Works Committee meetings; for, as far as he could recall, the Committee used to deal with far more important topics than had been the case during the past year.

Did they not then agree, asked the consultant, that it was a striking fact that they had spent so much time considering how to train themselves in management subjects, and so little in comparison on how to maintain contact with their constituents? They might wish to consider the potential training value of making and listening to departmental reports, and of culling out from such reports that material which would give them a picture of what was going on in the works as a whole.

The Works Committee Secretary agreed that the purpose of the Committee could not be very clear, for, as far as he was concerned, he really did not know exactly what his own duties were, and this could not have happened if the Committee had had a definite purpose. It was not that he wanted very detailed specifications for his job, but he would like to have some rather more clear guide-lines and, perhaps, if his role could be made more clear it might help to straighten out what they were all supposed to be doing as committee members.

Many of the others were disconcerted to find that their Secretary had no clear idea of what he should be doing; but the fact that it was not just a problem for him alone soon came out when different members began to express their own ideas about his job in order to help him. At one extreme there was the view that the Secretary might save everybody's time and see that a better job was done if the elected representatives could pass along to him all the problems which came their way; he could deal with these problems in a more uniform manner, perhaps assisted by on or two other full-time people. At the other extreme was the view of those, including the Secretary himself, who felt that the Works Committee members should handle all their own affairs without the Secretary being called in except when personality clashes or the "atmosphere" or "climate" of a department was such that someone outside the immediate situation was needed. The difficulty with any scheme, and one that was a constant source of trouble and irritation to him, added the Secretary, was that not everyone was equally courageous, experienced or willing to take the trouble to take care of his own show.

"The Committee seems to be in a state of despair. This despair, we feel, was reflected in the decision to hold meetings once a month. This decision, while possibly justifiable, gave the impression that the Committee was also indirectly giving expression to a wish to pack it in."

The Consultant emphasized that this paragraph was a highly impressionistic statement. He did not wish to imply that to cut their meetings down to one a month was either correct or incorrect; but the decision did seem to be partly tied to their being somewhat depressed about the worthwhileness of what they were accomplishing. Perhaps they recalled that he had mentioned to them in July, about six months before, his impression of their being bound down by a feeling of futility to the point where they wished to pack things in. That was at the time when he had reported to them the Research Team view that they would not finally settle the issue of the time and cost of consultative activities until they got a clearer idea of what they hoped to achieve by all their efforts.

The Service Department representative said he felt that these comments were absolutely correct. The Works Committee was not progressive enough in its thinking and activity. "We are working for the workers, not for pay." Could they claim that they were forward-looking? He thought not, for they had become bogged down in routine matters. They should be discussing what was going to happen to them in the next five years, and particularly attempting to decide what they wanted to happen so that they could plan constructively. If they themselves could not make up their minds what they wanted, then they could hardly expect to be looked upon as real leaders of the workers.

The Chairman concurred in the suggestion that one of their problems might well be their lack of known and agreed policy as to what they were trying to achieve. He was followed by the Secretary, who also said that it was certainly clear that they had to find out more specifically in what direction people wished to travel; the only trouble, he said, with a slight tinge of cynicism, was that no one ever put forward any of these progressive ideas. If they were to get anywhere, people would have to be a lot more forthcoming with their views. Perhaps, he concluded, it would be wise if they now turned to the tentative suggestions which ended the report, for in them might be found practical steps which they could take to do something about these plaguing troubles.

"The above comments are intended to indicate that one of the central problems of the Works Committee may be its own internal relationships. Until its role is clarified, and members of the committee know where they are going, and why, it is likely that the Committee will continue to be a block in the channel of communications between the Works Council and the shop floor. In order to lead to such a clarification, and possibly to improve the relations within the Committee, the following suggestions are made:

(a) that there is a need to discuss what members hope to contribute to and get out of the Works Committee, and to clarify the role of the Committee Secretary, and of the individual Committee members. This will require working through the problems of relations within the Committee, some of which were illustrated above.

(b) the Research Team would be willing to assist in these explorations of relationships within the Committee. Our recommendations would be:

(i) that one team member should sit in at Works Committee meetings to try and help clarify difficulties as they arise;

(ii) that he might meet with a subcommittee to consider in more detail points raised in this report and any other matters relating to the work of the Committee."

Both these recommendations were accepted; the arguments put forward in favour of having a Research Team consultant indicating to some extent the frame of mind of the Committee at that time. They could have somebody with no axe to grind commenting on what they were doing, and this would undoubtedly have training value for them. Since the consultant would not be a rival of anyone in the Committee, and would not be a member of any group, and since he was not tied to the company, he would be able to speak freely. And they did not, they thought, really have to ask what the consultant's role would be because they had already experienced it in the discussion of the report which they were convinced must have a speedy and constructive effect.

A subcommittee of six members was appointed to meet with the consultant and bring back further recommendations on how to improve communications. Much of the discussion had apparently struck quite deeply, for there was general support for the hope expressed by the Tool Room representative that the subcommittee should do its job and report back quickly, for they had little time to lose.

UNRESOLVED PROBLEMS OF COMMUNICATIONS

On 21st February, 1950, the subcommittee on channels of communication brought back its report. In an introductory comment, the B.2 department representative, who had been the convener of the subcommittee, explained that they had discussed the relations among representatives, the general apathy in the factory and among representatives, and the fact that no one knew what was going on in any one else's department. They had decided that the first step they had to take was to straighten out their reporting problems, and so they specifically wished to recommend:

(a) each shop committee should give a monthly report to the Works Committee Secretary on the most important events happening in the department:

(b) the Works Committee Secretary should then digest these

reports, add his own comments based on his experiences
during the month, and bring back to the committee a comprehensive report which would integrate and pull together
all of the main events of concern to the workers occurring
in the factory during the month;

(c) the Works Committee to discuss this report so that they
could determine which of the issues reported upon were
of most immediate and widespread importance, in the
sense that they were worrying the workers and had to have
some action taken;

(d) on the basis of these discussions, the Works Committee
would then report to Works Council those issues of
central importance to the workers in the factory, in the
same way as the Divisional Managers reported, so that
appropriate steps could be taken to settle any problems.

Delaying Change

The recommendations of the subcommittee on communications were unanimously carried. But they could not be implemented immediately, for the Works Committee, at the suggestion of its Chairman, had decided to incorporate the new procedures into their constitution, presumably to make absolutely
sure that they would be carried out without fail. This required
that the proposals be drafted in a form appropriate for the constitution—a task which was assigned to the executive subcommittee.

At the next meeting, on 21st March, the Chairman reported
for the executive subcommittee that they had been unable to
tackle the re-phrasing of the proposals for the constitution
because of other pressures, and so the matter would have to be
delayed a further month. He also took the occasion to remind the
Works Committee members that they had delegated to the
executive subcommittee another and related job; that of drawing
up proposals for training a special panel on management responsibilities and methods. In view of their recent deliberations,
he now proposed that this idea should be dropped and that they
should get on with their duties of becoming efficient workers'
representatvies and not mere shadows of management.

The Chairman's proposal was adopted with little comment
and the meeting went on. But they immediately ran into the very
problem of not being able to put forward issues of serious

import from the shop floor. The fourth annual Works Conference was scheduled to take place in two months' time, in May. The Personnel Division, which was responsible for organizing the Conference, had asked the Works Committee, along with other representative bodies in the factory, for their views on how the Conference might be run. Advice was also sought on what were currently the matters most vitally affecting their constituents which could usefully be brought out into the open and discussed at the Conference, so that action beneficial to them all could be initiated. The Committee considered this request, talked about the need for having plenty of opportunity for small group discussions, but failed to bring forward any suggestions about things bothering their constituents, in spite of the fact that the Works Conference was being organized as an opportunity of getting something done about things which had gone wrong.

The failure to put together any proposals for subjects for the Conference was particularly striking. It had been the Works Committee itself which a few months before had been instrumental in getting accepted the policy of making the conference into a working event which would get down to serious discussions about factory problems. With this in mind, the consultant, who was now sitting with them in his new role, commented that here was an example of a practical communications difficulty. They obviously regarded the Works Conference as an opportunity to raise things that were distressing to the workers; and yet no topics had been brought forward to be sent to the Personnel Division for inclusion in the conference agenda.

The Chairman tried to stimulate discussion by giving examples of the sorts of complaint they had talked about on different occasions, which might usefully be raised, such as the problem of scrap, or of the recording of time-keeping since the ending of clocking-on. But not much response was forthcoming. The Secretary, and the Tool Room and Service Department representatives also tried to stimulate discussion, for here was a chance to do something to better the lot of the workers. But still there was no lively interest from the others; and even those members who were among the real leaders of the Committee were themselves unable to do what they were exhorting the others to do, to bring forward any ideas about topics or problems which required taking up.

Avoiding Change

Despite the upsurge of enthusiasm for straightening out their problems of two-way communications with the shop floor, the difficulties of the Works Committee in representing the views and attitudes of their constituents in an effective and telling way went on. The proposals for reporting from shops were not incorporated into the constitution because, within a few weeks of the above meeting, fundamental changes in the composition of the Committee were mooted in connection with discussions which had been proceeding between management and the district trade union officials. It was thought to be the wisest course to leave any changes in abeyance until the possible new developments were known. While there was much to be said in favour of such a course of action, this decision also contained elements of a wish to avoid implementing the new methods of reporting put forward by their subcommittee—methods which they had already adopted in principle.

The notion that there was some stalling was given added strength by the degree of concern expressed by at least two members of the Committee, who felt matters were being unnecessarily shelved. The B.2 representative, who had been the convener of the subcommittee, did not see why the proposals, which were consistent with trade union practices, could not be implemented right away. Even if they became a shop stewards committee, they might just as well start at once with reporting from shops. The Tool Room representative, who was a member of the Internal Development Committee, had helped plan the Works Council communications project, and he argued that if they were to continue to put things off in this way, then there was a danger that the Works Committee would get into the position in which, rather than playing a leading role, it was likely to act as a barrier to progress in the factory. The matter was settled, however by the Chairman. He pointed out that not only was there nothing to prevent any member from reporting anything of importance from his shop, but that their discussions had revealed that there was every encouragement that this should be done. They were only putting off for the moment the setting up of formal machinery for reporting.

Although the consultant did not take it up at the time, he and the other Research Team members were left with the strong impression that members of the Committee had regressed to their previous attitudes towards communications. A hardening of resistance towards any new methods had set in. But what lay at the root of this resistance they were unable to detect.

One thing, however, had become increasingly apparent: the Works Committee members were working under very great strain. Because they were so anxious and uncertain about their prestige both in the eyes of their supervisors and of their fellow workers, many of them experienced acute discomfort and even fear at the thought of approaching workers or supervisors in their shops, as was essential on occasion if they were to carry out to their own satisfaction their duties as representatives. In a general way, this dread of facing the dis-interest of others could be said to underlie much of the resistance to the setting up of two-way communications. For it was this very communication function which most threatened to force them into contact with their constituents. The problem of communications has been left for the Shop Stewards Committee to tackle.

WORKERS' LEADERSHIP UNDER CONDITIONS OF CO-OPERATION

To understand and appreciate the difficulties in the position of the Works Committee it is necessary to turn first to certain general considerations. Throughout the war, and afterwards, workers in the community surrounding Glacier have been able to gain more or less constant employment in the local industries. Even when redundancy hit Glacier, along with the other light engineering firms in the district, 80 per cent. of those released were able to obtain new jobs before they left. This comparatively secure employment situation has been supplemented inside the factory by the record of Glacier's management-employee relationships and by the social developments which have been in advance of demands by employees. As a result of the local employment situation and the favourable attitude towards the management, two of the sources of disturbance which usually facilitate cohesive workers' organization have been partially removed. Indeed the desirability of employment with the firm has led workers to look to the management rather than to the trade unions for security of employment, and has aroused in the workers' leaders an acute conflict over loyalties divided between the firm and the trade unions.

In view of the fact that the district trade union organizers have themselves emphasized the need for the workers' leaders at

Glacier to adopt a constructive and co-operative attitude towards the current developments, the discomfort experienced by the Glacier leaders in their efforts to maintain a strong organization cannot easily be explained in terms of divided loyalties only. The fact that the Works Committee's discomfort eases when there is an issue to fight about, and increases as the situation becomes more settled and calm, would suggest that the work of the committee has called forth a high proportion of leaders who function most easily and effectively in a fight situation. As was pointed out in a report to the Works Council, the Council members felt that they were not really demonstrating leadership unless fighting for or against something. Many of the Works Committee members are not so much concerned with carrying on a fight for its own sake as with fighting specifically on behalf of others; they derive satisfaction from being able to do things for them, protect their interests, and see to their welfare. The reasons for the satisfaction are undoubtedly complex, but in general it can be observed that being able to do things for others brings satisfaction for needs to nurture and protect dependants, as well as for needs to control others by making them dependent.

If these assumptions are correct about the needs of these workers, leaders—of some to control and to protect their constituents; of others to fight on behalf of their constituents; and others to do both—then it is possible to understand rather more clearly what happens at times when there is an issue to be fought about, and at those times when there is not. When there is a fight on, all of the needs of the elected representatives can be satisfied. The constituent gets his protection if he allows himself to be dependent on the representative he elects. The representative derives the satisfaction of having people dependent on him to take care of their interests. Because a fight is called for, aggressiveness in behaviour becomes a quality which can not only be freely expressed but whose expression leads to approval. The elected leader can also turn his aggression against his own constituents and get back at them for their previous disregard of him; "I told you so", is a commonly held attitude at the beginning of a crisis.

When affairs are more or less steady, the situation is exactly reversed. Constituents become uninterested in their representatives and set up barriers to shut themselves off from the responsibilities of supporting the consultative structure. Jobs, home life, and other activities outside the factory become the centre of interest.

The representatives, in the face of what amounts to loss of their followers,become disconcerted and angry with their fellow workers whom they criticize as apathetic. These feelings are caused in some by the frustration at losing a group of dependants to care for and control, and in others by the loss of a setting which not only permits but sets a premium upon the expression of aggression. In either case, getting angry with their constituents makes matters worse, because it leads to the fear that they themselves will be made the target of anger in return. Whether or not this actually happens does not matter. The fear of anger is enough to increase the sensitivity of the leader to criticism and. to feelings of being rejected.

Returning to the problem of the occasional despair and sense of futility experienced by the Works Committee, it may be that a partial explanation lies in the recurring loss of their dependants and followers, with the consequent thwarting of their wish to lead and to control by leading. They have to work side by side with the very constituents who have for the time being rejected them, and they are thus faced in their everyday work with the fact that they are groupless leaders. This gives rise to annoyance and anger which cannot easily be expressed directly, and most of which must then be bottled up inside. Such a mixture of frustration and un-expressed annoyance is quite sufficient to cause despair and futility, particularly when one's offers to do good are turned down. The saving feature of the situation is that representatives may be called upon at any time for assistance in individual appeal cases. But even here the self-respect so gained leads to satisfaction only for the small number picked for this role, the other members of shop committees and Works Committees being used merely to give moral support.

In effect, life for many representatives consists of being the leader of a group of people who at one moment demand his leadership, and in the next moment reject and scorn him in order to get on with their work and earn enough to keep up their private interests and responsibilities. Small wonder then that the representatives will often seek to discover or create new problems, or else look around for support from, and alliances with, other groups, rather than tolerate the sense of isolation brought about by inactivity. The management group at Glacier, with its interest in joint consultation, constituted a natural potential ally in these circumstances, and the workers' leaders have from time to time derived their main support and leadership from like-minded members of the management, who in their turn also were seeking relationships outside the responsibilities of executive leadership.

Delegated and Representative Leadership

Another way of looking at the function of the elected leader is to consider his position as carrying various degrees of delegated and representative authority; a *delegate* being bound by the specific instructions he receives from his constituents, and a *representative* being given the authority, within broad limits, to use his own initiative in coming to decisions. Although it is laid down that the Glacier elected leader shall be a representative, nevertheless, in practice, he may often have only delegated authority when his constituents are vitally interested in a given issue and instruct him continuously about their wishes. Frequently, under such conditions, there can be observed a banding together of the leader with his constituents, with a high degree of cohesion, inter-dependence, and freedom of communication; these factors, as well as the commonality of outlook and purpose, make it possible for the leader accurately to present the views of his followers. Trouble arises in that behaviour as a delegate usually irritates the other people with whom the leader is carrying on negotiations, for delegated authority means a minimum of flexibility in discussion, and allows the stalling device of constant reference back to constituents. To have delegated authority is most satisfying for the elected leader, in that it provides closest contact with his constituents and most immediately experienced leadership and control of a group.

More difficult for many members is to carry the duties of representative. The usual situation is that there are no immediate issues which capture the direct and active interest of the main body of employees. The elected leader is therefore left to his own devices, to represent the views of constituents as he himself sees fit. Many, however, are only in a position to approximate to the views of their constituents. There is little that most of them can do about it because they feel rejected and pushed out. And if they try to do anything, they run up against the barriers which their constituents have erected to prevent anyone from breaking in upon and disturbing their withdrawal from participation in the affairs of the factory, outside their own immediate sphere of work. It is just at this point that so many members of the Works Committee become despondent.

All but the more experienced have been unable until recently to perceive two highly constructive roles which are open to them: namely, that of maintaining an efficient committee, and that of accepting the role of representative, exerting leadership in such a way that their legislative record may speak for itself. With regard to the role of maintaining an efficient structure, a change in attitude has taken place as a result of the discussions about changing from the existing set-up to one based on a shop stewards structure. Through these discussions and the recent ballot in favour of a shop stewards set-up, all of the workers' leaders have been faced with the serious implications of structure building, and this has tended to strengthen their perception of their function as a "just in case" group, or core group, with the task of maintaining an organization which can go rapidly into action and be modified so as to meet any emergency. If such a role is accepted, the positive values in maintaining a club-like atmosphere during the phases of representative authority become apparent. For this kind of atmosphere makes it possible to work through stresses in the group, and generally to keep members up to scratch, even though there may be no urgent business on hand at the time.

With the change to a shop stewards set-up the volunteer, part-time elected leaders' group becomes backed up by a permanent organization—the trade union structure with its full time paid officials. The interplay between the part-time volunteer and full-time official can make possible the nurturing of the workers' organization inside the factory by means of the outside strength, guidance, discipline, and skill which the trade union movement can provide. But against this, there are already indications that conflict may arise out of the greater inflexibility brought about by some factory leaders gearing their policies to what they believe will gain them the approval of outside trade unionists rather than to the desires of those who elect them. The district union officials have attempted to discourage such an outlook, and have asserted time and again that they neither proposed nor had the time to make decisions for the Glacier shop stewards. The problem, however, does not end there. The very fact that the district officials are more knowledgeable on organizational matters discourages initiative in the local leaders and leads them to seek official approval before acting. This was seen in the case of the leading role which the officials accepted in the discussions with the management which led to the new set-up. Failure to clarify these unresolved problems of the relations between the district officials and the local leaders, along with the unresolved differences between various trade union cliques, had a great deal to do with the narrow margin of the vote in favour of the shop stewards set-up. This vote indicated fairly widespread doubts among the workers whether the trade unionists would provide the best consultative leadership.

There are indications, however, that a change is at hand. The changes in outlook taking place in the executive structure of the factory, and the creation of a sharp separation of executive from consultative work, has led the management members to make greater demands on elected leaders. Management proposals for new policy or for changes in policy are brought to the Works Council. This makes demands on the elected representatives to discuss, modify, and finally ratify policy. This procedure is in contrast to the semi-collusive attitude which was previously so strong, in which the top management and the elected workers banded together against the rest of the factory. In fact, top management is now taking as its primary responsibility the operation of an efficient executive, and is leaving to the elected representatives the full responsibility for looking after their own affairs. This change in attitude has shown itself in a number of instances recently where the workers' representatives have had to rely on their own resources in taking action, where previously they could have looked for management advice and support—useful at the time, but a doubtful benefit in the long run.

Whether or not the consultative mechanism will serve to diminish these problems of the intermittent satisfaction of the elected leader, it is hard to say. It seems reasonable to suppose, however, that by making greater demands on elected representatives to give advice about the attitudes of employees, so that effective policy can be framed, the new set-up will give a more positive function to representatives than is the case where the consultative machine is used only as a trouble-shooting mechanism. In addition, the recent clarification of the functions of the Works Council, and hence of the Works Committee, and the restructuring of the Council, will both contribute towards creating a more definite and secure place for the Works and Staff Committees in the scheme of things.

CHAPTER SEVEN

THE SUPERINTENDENTS COMMITTEE

Organizational Problems in the Management Chain

TOWARDS the end of April 1949 the Research Team was requested by the Superintendents Committee to discuss with them possible changes in the constitution of the committee. This relationship with the superintendents, which began as assistance to them in discussing what on the surface seemed a minor organizational problem, soon turned into a joint exploration of how group problems may stand in the way of effective organization, and of how the initiation of joint consultation between higher management and workers' representatives may have disturbing effects on the operation of the executive system, if due regard is not given to relations in the executive chain.

COMPOSITION OF THE COMMITTEE

The Superintendents Committee is a body of seventeen persons, originally formed in 1940 as a "Foremen's Committee", to provide the shop foremen with the opportunity to discuss shop problems among themselves and with the Works Manager or the Works Director. Gradually, however, the composition and function of the Committee changed. Others, outside the Works Division but at superintendent level, were co-opted for special purposes, and the Committee changed, to comprise the seventeen people shown on the accompanying organizational chart. It will be seen from this chart that ten were Works Division Superintendents, directly or indirectly under the command of the Works Manager. One, the Chief Inspector, was at the same level as the Works Manager, and another, also of Works Division, was from the Production Engineering Department under the Chief Production Engineer. In addition to these ten, there were the Training Superintendent from

185

Personnel Division, the Shop Superintendent from Service Division, and the Stores Controller from Finance Division. It will be seen that the Committee did not correspond to any existing groupings on the organizational chart.

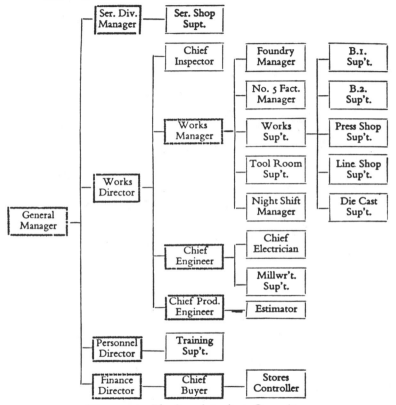

Diagram 13. The Superintendents Committee.

Members are shown in solid blocks. It will be noted that the seventeen members come from seven different departments in the four largest divisions (the Technical, Medical, and Commercial divisions not being represented).

STATEMENT OF THE PROBLEM

The consultant first met the Committee on the afternoon of 11th May.   The Chairman, opening the discussion, explained that they had two main problems.   The first was a feeling that their business was not as efficient as it might be because they did not have sufficient top management guidance.   The second, that they spent a great deal of time discussing things, but seemed unable to arrive at satisfactory agreement. Other members picked up the first point, some maintaining that it was essential to have a higher management chairman to direct their discussion along the most efficient line and keep them within the limits of management policy, while others felt they should continue with their own so-called democratic meeting with an elected chairman.

Noting that they were decidedly lacking in agreement at that moment, the consultant suggested that what the Superintendents were really seeking was a higher management representative to make them agree. This comment was received with smiles, and they went on to discuss whether full agreement was always necessary, some holding that there were many issues on which full agreement was not necessary so long as they each knew what the others were doing. But they were unable to agree on this point either.

The consultant commented that their behaviour supported the Chairman who had said that one of their problems was this difficulty in arriving at agreement with each other—a statement which had been followed by fairly unanimous headshaking. But there was experience to show that when committees of this kind were unable to arrive at agreement, there were often unrecognized difficulties in the group which prevented effective discussion. This occurred because the unrecognized issues tended to creep, unperceived, into the discussion, so that, although the argument might seemingly refer to the issues at hand, the emotional content of the discussion, in contrast to this, would be more closely related to the unrecognized problems. This mixing of emotional undercurrents which were unrelated to the matters consciously under discussion led to confusion and frustration.

At this point, one of the committee members interrupted to deny that they had any personal differences in their group, and others strongly agreed with him. The consultant hastened to explain that he was not only referring to personal differences; but he had the definite impression from the tone of their discussion that there were unrecognized difficulties of some kind which hindered them in their work. He could not suggest any easy solutions, and in order to get the feel of the situation it would probably be necessary to attend a series of meetings. If they wished such help, it could be given, but they were warned that no results could be guaranteed. He then left, suggesting they might talk over whether they wished for this rather more extended collaboration.

A LACK OF POLICY

The Superintendents having decided in favour of collaboration, the consultant began on 8th May to attend their weekly meetings. A number of difficulties soon became apparent. One of the most immediately striking features was the difficulty in starting. Nine members constituted a quorum. The meeting was scheduled to begin at 3.30 p.m. It was very rare, however, that the required quorum of nine had turned up before 4 o'clock. Until this time, the others just sat around having informal conversation. About once in every three meetings, however, there was still no quorum by 4 o'clock; the Chairman would therefore call the meeting off.

There was also the problem which was encountered not only among superintendents but in the Works Council, Divisional Managers Meeting, and in other sections of the company. This was the vexed question, "what is company policy?" The matter arose on 8th June, during a discussion of the responsibilities of their immediate subordinates. Reasoning that it was their own responsibility to implement company policy in their own departments, they decided that it must be the responsibility of their subordinates to implement departmental policy in their sections. This all seemed very logically consistent and neat, until one of the superintendents asked, "What is this policy which we are implementing and asking our subordinates to implement?" The others were momentarily crestfallen, but soon recovered and got into an argument over whether there was any clearly formulated company policy, some roundly maintaining there was, and others just as vigorously protesting there was not.

But, agree or not, there was no one in the room who was able even to make a first approach to stating what company policy was; about the best attempt being that of one superintendent, who explained, "Well, it's hard to say exactly what company policy is, but we all know inside us what it is, for when you don't follow it you soon have higher management down on you!"

They finally dropped the point by agreeing that possibly at some future date they would ask someone from higher management to come down and give them a dissertation on company policy. Little did they realize that at that time higher management members would have found the task no easier.

A month later this question of policy came up again by an indirect route. The superindendents' member of the Works Council in reporting on the Council meeting of 22nd June, had been describing the Research Team Project Officer's references to the lack of clarity in distinguishing executive and consultative functions. The other superintendents seized on this point. They all felt they were suffering to some extent because the respective responsibilities and relative authority of supervision and shop committees were not clear. One member expressed the general mood when he proposed that what they needed was a clear statement of company policy from management. The others warmly agreed. But then they all suddenly laughed when they realized that this was just what they had not been able to agree upon a month before.

Looking back, it seems likely that the Superintendents' intense concern at the absence of policy was a reflection of their feeling of having been abandoned by higher management. In a sense, their committee itself could be regarded as a banding together of middle-layer management, from various Divisions, in search of status and security which was not provided by their senior executives in their own Divisions. But why, we must ask, were these members of the middle management group unable to seek a solution of their difficulty from the senior executives in the various divisions concerned? This brings us to certain unexpressed difficulties among the superintendents themselves, which stood in the way of effective action.

RELATIONS AMONG THE SUPERINTENDENTS

The consultant, feeling that he had become sufficiently familiar with the general atmosphere and attitudes in the meeting, suggested, towards the end of June, that he might help tackle their problem more directly. He proposed that he should arrange for interviews with each of them individually, to take up with them their views on the organization of the Superintendents Committee, its functions, and its value to them. After having conducted these interviews, the material would be put together in the form of a report to them, which would not refer to individuals but would sum up the general feelings of the Committee as a whole. This proposal was adopted, and the following material is a digest of the report which was drawn up.

Findings from Interviews

Most of the Committee felt that the agenda for their meetings did not contain sufficient matters of general interest to them all, and there was always the lingering doubt whether you might not be better off attending to your own job. Along with this was the almost unanimous wish of the superintendents from the Works Division to have a regular meeting with one of the higher executives from their own Division. They felt isolated and out of contact with higher management policy, about which they were not clear, but which they were supposed to implement. The earlier form of the Committee, when it was a meeting of shop foremen only, was held to be more satisfactory because they were a group with a common interest.

The majority of the superintendents did not consider that they participated in formulating company policy. They were less effective as individuals, and as a group, because higher management did not take them into its confidence in explaining its long-term aims and objectives. An example of this was the fact that when the Works Superintendent had recently retired, management had not discussed with them the new situation or future plans.

Nearly all the Committee felt that the weak link between themselves and higher management demonstrated itself in the difficult position of their representative on the Works Council. He could only speak or vote as an individual, because of the difficulty in briefing him, due to their lack of knowledge of company policy; and, in any case, there was little that he could say, surrounded, as he was at Council meetings, by practically the whole of top management.

Taking these findings from the interviews along with the material gleaned from attendance at meetings, it became clear that the problem whether they should elect a chairman from among themselves or get someone from higher management, was only one symptom of a much greater difficulty; for, with only one exception, none of the members of the committee had formal and regular two-way consultation with the higher management members in their own divisions. The first problem to be solved, therefore, appeared to be that of co-ordination of activities within divisions. In view of this, the consultant suggested that the members of each division should take up independently the question of divisional meetings to serve the purpose of regular two-way communications within the executive chains on policy and executive matters.

But, even if the superintendents were to agree that this suggestion had any merit, would they necessarily be able to carry it out? The consultant's assessment was in the negative, for apart from other factors outside the committee, there were certain difficulties in their own relations with each other which were making them less effective as a group. Apart from purely personal differences, these consisted mainly of a split between the "practical" men, or those who had come up the hard way from the shop floor, the "technical" men, or those younger men who had come in from technical colleges, and the "administrators", or those who specialized in administrative rather than engineering skills.

Over a period of years, higher management had vacillated in the matter of the prestige it attached to these various skills, but latterly the belief had taken hold that all that was required to get ahead was to be able to talk about human relations and be facile at "paper work". Knowledge about making bearings was supposed to be a secondary consideration. Whether or not they were correct, these beliefs nevertheless existed, and created not inconsiderable tensions based on status and prestige deriving from the possession of certain skills. Such tensions were stated to exist among the superintendents, exacerbated by the fact that some of the younger members had been given Grade I Staff status, in contrast to the rest who remained in Grade II.

The consultant's report ended with the following proposals:

> *First, straighten out the matter of two-way communication and co-ordination within Works Division, and possibly some other Divisions, particularly on matters of principle.*
>
> *Secondly, consider whether it seems desirable to change the present Superintendents Committee into a truly inter-divisional conference, or conferences, to discuss general policy matters affecting the company as a whole.*
>
> *In short, the proposals consist in removing the present business of the Superintendents Committee concerned with Works Division*

*alone, into Works Division meetings; and in enlarging the scope
of the more general business to the level of overall policy discussions
concerning all divisions.*

In making these proposals, he emphasized that success or failure
would probably very much depend on the degree to which they
were able to sort out their own group relations, in order to be
able to take effective and agreed action.

Discussion of the Consultant's Report

The report was discussed at the committee meeting on 12th
July. The Chairman stated that he thought it was a very comprehensive statement, and others commented that the report
did cover the situation, and that the final proposals particularly
should be voted upon and agreed. Accordingly it was moved
that the summary of proposals at the end of the document be
adopted. This was unanimously passed.

Having thus disposed of the report in a general way, they
then went back to the beginning in order to go through it
paragraph by paragraph; but they did not get very far, for by
the time they had considered the first few paragraphs, many
conflicting points of view had been expressed. Thus, for example,
they talked about the general impression gained by the Research
Team that many of the topics discussed in their meetings did
not arise from the immediate felt needs of the meeting as a
whole, but only from the interests of a few people. One of
them pointed out that, when the meetings used to be foremen's meetings, they had had very good discussions about
technical subjects. One of the others, however, argued that
this was not enough, and that there were some members of the
committee who did not take a sufficiently active part in discussions about administrative and policy matters. To this the
reply was that although some might not be too comfortable
discussing more general matters, they could nevertheless run
their shops, and run them efficiently.

At this the meeting fell silent for a moment—a silence broken
by one superintendent, who pointed out that they seemed to
have got on to what he felt was one of the most important
points raised in the report; namely, the differences among the
superintendents themselves. There were some of them who
had come from technical colleges, and did not have all the

practical shop experience possessed by some of the others who had come up the hard way. He knew that these superintendents were rather disparagingly referred to by the others as "paper boys", and he thought it was high time that these things were straightened out. Another superintendent reminded them that it had been the general feeling for a number of years that administrative skills were valued more highly than engineering skills—as it had often been put, it was the "paper boys" who were the "white-haired" or "blue-eyed boys" of higher management.

At this another member of the committee said that he knew they sometimes talked about each other outside meetings using terms about each other such as "blue-eyed boy" and "paper boy". He felt their present discussion was useful, because it was better to get these things out on the table. From this point of view they would all have been better off if they had had such a discussion three or four years ago. If they had, they would not be having their present trouble with the organization of their committee.

The consultant commented that they had discussed a most important part of the report. This seemed to be just the kind of underlying issue which may well have contributed to their difficulties in getting agreement at their meetings. There had been much feeling expressed in the preceding discussion and, as one of them had put it, this had been going on for about three or four years. This would suggest that the emotions surrounding this particular type of problem were present in many of their discussions and prevented them from reaching effective solutions.

WORKING OUT A NEW CONSTITUTION

There was a noticeable diminution of strain and tension in the Committee at its next meeting on 20th July. Everybody seemed cheerful. It was the most strikingly light-hearted atmosphere the consultant had seen during the two and a half months he had been meeting with them.

There was no reference back to their discussion of the week before; their attention being focused on the possible setting up of a meeting of Works Division superintendents. This raised a difficult problem for the group, for although there were seventeen members of the committee, only about six to ten were anything like regular attenders and, of these, three were from outside the Works Division. If, therefore, they decided to have meetings of the Works Division alone, the three regularly attending outsiders would have to be excluded, and many of them were loath to see their present group split in this way.

Thus, although intellectually they were agreed that it would be more effective to hold separate divisional meetings, nevertheless they could not face breaking up their existing interdivisional group. In such a situation it seemed not unreasonable to suggest to them that they were emotionally banding together against some outside individuals or groups which they felt to be unsympathetic, or even hostile, towards them. This interpretation, however, was apparently disregarded at the time.

In discussing a possible Works Division meeting it soon emerged that they felt there was little opportunity for two-way communications with the Divisional Managers. It was this which partly caused their confusion about company policy, for, in effect, they had little opportunity to participate in the making of policy, nor did they have the opportunity to hear regularly the results of deliberations by higher management. This was true for the other Divisions as well as the Works Division.

But the story was not completely one-sided. There was considerable room for doubt as to whether they preferred to take part in policy making, or to receive clear-cut instructions without having anything at all to do with the policy on which these orders were based. These mixed feelings contributed to the already complex problems which faced them, and few were really certain whether they wished to change the form of the meeting. As they were, they could carry along their difficulties with a minimum of interference from outside. If they changed, they were faced with the uncertainty whether divisional meetings in Works Division and the other divisions concerned would be satisfactory. They were frightened of the isolation which threatened if these divisional meetings failed.

Resistance to Change

As members of the Committee began to realize that the discontinuance of their meeting as it was then would mean that they would have to rely on divisional meetings, that is, on being able to establish more satisfactory relationships with their Divisional Managers or their immediate superiors, some of them began to feel more strongly that perhaps it would be best to keep the present meeting as it was. Or, if there was to be a change, then should not some appropriate interdivisional meeting be created as well? This request came strongly from those who felt that, in the existing Superintendents Committee, members from various divisions could support each other, whereas once they lost this committee they would have to rely on their own abilities to handle situations in their own divisions without support from the others.

Because of these unresolved doubts, final action was held over till the next meeting when, it was agreed, members should bring proposals to the group. In response to this decision, one of the members brought forward the following proposition, which became the main focus of discussion:

> "Regular meetings of superintendents and their chiefs should be set up within each division where such meetings are not already being held. . . . Until direct two-way communication is firmly established in each division there is not a general case for an inter-divisional conference. . . . When two-way communication is firmly established, the setting up of an inter-divisional conference raises issues beyond the power of the present Superintendents Group, and higher management should be informed of this with a view to opening discussions regarding an inter-divisional conference."

Discussion of this proposition brought up once again all the old difficulties in the committee. There was a split. Three superintendents were in favour of the above resolution; three were in favour of a slightly modified version which laid greater stress on the need for an inter-divisional meeting; and the rest were in favour of both, feeling there was really no difference. Because a 75 per cent. majority was required they were unable to arrive at any agreement, and a stalemate resulted, since neither of the groups of three would give in to the other. It was pointed out that this split seemed to be partly based on the "practical"—"technical" difference in the group. This point had, indeed, emerged fairly obviously for, as one of the members put it, referring to the opposing group, "If you don't vote for our propositions, I don't see why we should vote for yours".

The difference being unresolved, the discussion had to be left over for yet another week, when, as a way out, a further resolution was put forward, the main purpose of which was to set up two committees: one, a meeting of the Works Superintendents alone, the other a meeting of the existing Superintendents Committees extended by the addition of superintendent level representatives from Technical and Commercial Divisions to complete the inter-divisional set-up. But, like its predecessor, this resolution did not heal the split. To resolve their problem, one of the members proposed that they should adopt one part of the proposal, namely, to set up a Works Division superintendents group to meet every fortnight, and to include the other production superintendent from Service Department because so much of their work was in common. Other members from other divisions might be invited from time to time to these meetings.

This proposal was finally adopted. But it left unsettled what was to happen about an inter-divisional meeting, and a number of the Works Division members still felt somewhat uneasy lest they were leaving some of the superintendents from other divisions in the lurch. Finding themselves unable to resolve this problem, they set up a subcommittee to consider it and bring back a report before the end of September. In the meantime they decided to go ahead and implement the proposal to have a Works Division meeting.

As a result, however, of their doubts about the adequacy of these measures, Works Division meetings were held up. But the subcommittee, working rapidly, brought back about the middle of September the following proposals which, it will be seen, represented very much a compromise solution to the whole problem. They suggested setting up a. . . .

"Superintendents Group . . . to foster and further efficiency within the company . . . and to form a link in the executive channel of communications . . . a meeting to be held each week.

"Membership would be open to those in charge of production departments of the London factories, and include the Works Manager; and those in charge of Millwrights, Toolroom, and Electricians.

"In addition, invitations to send accredited representatives would be sent to Inspection, Production Control, Production Engineering, and Drawing Office, and each other Division apart from the Works and Service Divisions. The agenda

The Superintendents Committee—1949

for each meeting would be sent to the accredited representatives
before each meeting, so that they could decide if the department or division should be represented at the meeting and,
if so, arrange for the most suitable person from that department
or division to attend."

The main feature of this resolution was that it set up what was
essentially a Works Division Superintendents Group but made
provision for all other divisions to be represented if they wished.
Or in other words, a way was open for the excluded members
of the existing committee to come back again if they felt too
isolated.

At this same time the Works Council sub-project on communications had got under way. This meant that the channels
of communication in all divisions were to be looked into, and
it is quite possible that this had a reassuring effect on some of the
non-Works Division members of the Superintendents Committee.
Whatever the forces at work the above constitution was rapidly
adopted on 21st September. The new constitution settled, the
consultant asked whether his collaboration was felt any longer to
be necessary. Some members held that it would be helpful particularly in the next phase of their work to have continuing collaboration, with reports on the way their new set-up was developing.
This was agreed to.

THE RESULTS OF THE CHANGE IN CONSTITUTION

The new Superintendents Group came into being during
November 1949. The change in constitution, being essentially a
compromise, had not fully resolved the troubles of the middle
management stratum. The problem of two-way communications
in the executive chain within Divisions was partly to be met by
superintendents seeking meetings with their own superiors, but
some of the difficulties in this can be seen by examining what
happened within the Works Division.

The revised structure of the Superintendents Group is shown
in Diagram Fourteen. It will be noted that its thirteen members
were the Works Manager and his nine immediate subordinates,
together with the two subordinates of the Chief Engineer, and
the subordinate of the Service Division Manager. In other words
the structure of the Group was neither in line with executive re-

quirements—for it did not have an executive leader—nor well adapted for a consultative meeting. As a result, the new group did not have a very successful history, and the frustration of realizing that they were still not achieving satisfactory two-way executive relationships remained painfully close to the surface.

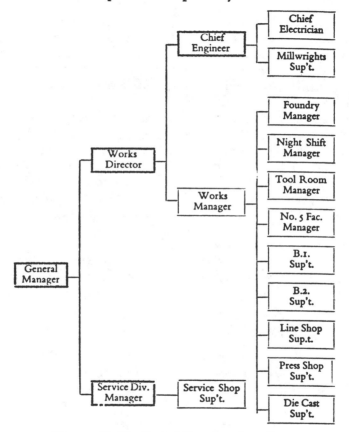

Diagram 14. The Revised Superintendents Group.
Members of the Group are shown in solid blocks.

The chairmanship problem was the first to be met. An attempt to resolve it was made by offering the chair to the Works Manager, as the most senior person at the meeting; but he declined on the grounds that he could not always guarantee to be present. This move having been unsuccessful, two members of the group were elected, one as chairman and another as secretary—a result which they viewed with no great enthusiasm since they regarded it as no great honour but rather as an undesirable chore imposed on the two most willing pairs of hands.

The formalities over, the new Group settled down to work. In considering the results, it is possible to leave aside that part of the constitution which called for the circulation of the agenda in advance of the meetings to all Divisions, so that other non-voting representatives could attend. This mechanism was largely a compromise designed on the one hand to reassure the excluded members and, on the other, to assuage the consciences of the Works Division members who had some feelings of guilt for wanting a meeting on their own. As things turned out, once the new Group was under way, the problems of holding effective meetings became dominant, and the earlier anxieties and feelings of guilt tended to lose their force, so that the mechanism for involving members of the other divisions never came into play. It was forgotten and faded away, unused.

The Functions of the New Group

The central business of the meeting was to consider regular reports by the Works Manager of decisions taken in Divisional Managers Meeting. In discussion of higher management decisions, there was some, even though slight, contact with policy matters. This dealing with policy also was reflected in an increasing tendency by the Group to call in other responsible managerial personnel to clarify various aspects of the running of the factory. In the course of the year's existence of the Group, discussions were held with the Works Director, the Chief Production Controller, the Chief Inspector, the Stores Controller, the Medical Officer, and others, on problems of implementing policy on such diverse matters as labour recruitment, time-keeping, sickness benefit payments, consumable supplies, and scrapped and lost materials.

In practice, however, all that the Group could do was to inform one or other of the higher management members of their views and difficulties, with no guarantee that anything would be done or that anyone would take up a case on their behalf. The fact was that, try to avoid it as they might, there was no one in the Group who was executively responsible for their activities and hence, in the final analysis, the most they could do was to pass resolutions, and hope something would happen.

On the consultative side the Group were little better off. They had their elected representative on the Works Council, through whom they could take up any matters they wished. They did not, however, once use this channel, and some of the reasons for this are fairly clear. The mode of operation on consultative matters was for the Works Director, on the morning before the Works Council meeting, to meet with the Group, as had been his custom with the previous committee, to discuss the Council agenda. This gave them the opportunity of expressing their point of view in a general way. But as they had had no chance to see the agenda beforehand, it was often difficult to discuss it intelligently.

But there was really a deeper difficulty than this in instructing the Group's Council representative. He was at Council as the only elected management member; all the others being higher managerial personnel appointed by the Managing Director. Who then was he, to speak the superintendents' mind in the face of such an array of his superiors? This feeling found open expression in March 1950, during discussions of the Internal Development Committee's special report on Works Council structure. The Group was much in favour of transferring its representation to the Grade I and Grade II Staff Committees, on the grounds, agreed by every superintendent who had ever been a member of Council, that the present position was untenable. Years before, one superintendent had gone counter to the general management position at a Council meeting, and had been "slapped down" so badly that it was ever after regarded as a most unwise thing to do. Since that time, the Superintendent's representatives had done little else than sit quietly by and follow the higher management lead in voting. There were also arguments in favour of maintaining their representation on Council; but we shall return to these below.

THE NEED FOR A CLEAR EXECUTIVE LINE

In short, then, the new Superintendent Group continued much of the old ineffectiveness of the previous committee. By becoming more firmly based in Works Division it had moved more towards an executive meeting function, and indeed had achieved somewhat more in the way of getting things done than had its predecessor. But still there was the feeling of it being a "talking shop", and this attitude showed itself in irregular and indifferently attended meetings.

One of the central problems remained incompletely resolved: there were still no effective arrangements within the executive structure for bringing the middle and lower ranks of management into two-way communication with higher management. This was particularly true at the mid-management level: the General Manager, at one end of the scale, did have regular meetings with his Divisional Managers, while the Superintendents themselves were supposed to meet their own subordinates at least once a month; but in the middle ranks, in nearly all the divisions, little in the way of formal two-way mechanisms existed. The provision of a so-called democratic committee and a place in the consultative structure did not meet the need for effective leadership in the executive chain. Joint consultation is not a substitute for management.

Some of the sense of isolation felt by the superintendents was reflected in the discussion of the Internal Development Committee's report on Works Council structure. In addition to the attitude mentioned above, that middle management would be best represented on Council through their Grade I or Grade II Staff Committees, there was also the fear that management would not be effectively represented on Council by the General Manager alone, a fear which in turn represented the sense that this group would be unable to have any effect on the General Manager's attitude.

THE EMERGENCE OF A WORKS MANAGERS EXECUTIVE MEETING

The confusion between executive and consultative work was brought to a head at a meeting on 4th July, 1950. The Works Director was going over the Works Council agenda prior to the Council meeting that night, and came upon a report prepared by the General Purposes Committee of the Council, on the subject of payment of workers' representatives for joint consultative work during periods when they would normally be working overtime. One of the Superintendents objected that this was an example of the Council going too far, and beginning to manage the detail instead of establishing principles. Was it not correct, he asked, that it should be for management to implement matters of detail of this kind, within a framework of Council policy?

Although agreeing in general with this view, some members

of the Group held that, since this particular matter had to do with payment for time spent on joint consultation (i.e. was a consultative matter), it should correctly be dealt with in the consultative and not the executive chain. The consultant, however, asked whether the day-to-day implementation of joint consultation was not in fact an executive matter. He had been struck, he said, by the extent to which, in practice, these joint consultative matters were split off from normal management responsibility. As one example of this, he instanced the case of the Superintendents Group itself, commenting that, in his view, their existence as a group was an attempt to overcome their felt lack of support from their higher management colleagues. This was a problem which he had pointed out in his original report to them, and one which had only partially been solved by their new structure.

The day previous to this meeting, there had been a change in the position of Works Manager; the Foundry Manager, who was a trained engineer, having been promoted to this post. The new Works Manager was most anxious to straighten out his responsibilities for his own subordinates. He was at this meeting, and concurring with the consultant's view, stated that in so far as his own subordinates were concerned he intended to meet regularly with them to deal with all matters related to the running of the Works, whether to do with technological problems, administration, or leadership. He felt that a tightening up generally within the executive chain was called for if the factory were to be an efficient producer and a satisfying place in which to work.

The new Works Manager's comments were endorsed by the meeting; one of the Superintendents adding that they had to get their own duties as executives right before they could straighten out the functions of the consultative machinery. The strongest support came from the chairman and secretary, who said that they had only that week spoken together about their function in the meeting, and had come to the conclusion that their positions were untenable because they had no executive authority in a group which in part dealt with executive matters.

A special meeting was called on 7th July, in order to discuss the future. The new Works Manager outlined his plans to bring together his own subordinates. Whether or not they wished to continue the Superintendents Group as such was not his affair, since it was not a meeting for which he could properly be held responsible. This they could decide for themselves.

In the ensuing discussion, the consultant tried to summarize his own views, arising from over a year's attendance at their meetings. Their needs, he thought, were for effective executive work. Without this their consultative and interdivisional co-ordinating attempts were doomed to failure. Indeed, did they not all now agree that these attempts to band together in inter-divisional groups were partly attempts to make up for their sense of isolation and lack of security in their managerial roles?

They had complained frequently, he continued, that the Works Council took decisions to which they, as superintendents, could not easily conform. But was it not partly their own failure as a group to agree what they did desire which prevented them from affecting company policy, and particularly from having any impact on the outlook of higher management? In connection with this, would they not concur that they often created the same problems for their subordinates as higher management did for them, in that many of them found it easier to meet with their Shop Committees than with their Section Supervisors? In other words they often used the elected consultative machinery in their shops in such a way as to by-pass their own subordinates in the executive chain.

Was there not some justification, therefore, for the view that the best contribution which they, as middle management, could make, not only to the company, but to their own comfort and well being, would be to take steps to ensure that higher management took note of their own views—and they would have to have views in order to do this. Then in their own departments they would have to give full support to their immediate executive subordinates, and straighten out their consultative relationships with their elected shop committees.

These observations by the consultant proved acceptable to the group, which, in an ensuing discussion of managerial roles and behaviour, endorsed the decision of the new Works Manager to call his subordinates together—in order to instruct them and to hear their views—in a regular meeting in which he would carry full responsibility both for the decisions taken and for raising with higher management the views of his subordinates. The Service Shop Manager, because of his common interests with this group, would be the only outsider invited to attend this meeting, and that only in the capacity of an observer; it being recognized as the responsibility of the Service Divisional Manager to brief his own subordinates on managerial policy and to make their views known to higher management.

With this decision the Works Manager retired, leaving the members of the Superintendents Group to decide whether they wished to keep their consultative and partially inter-divisional meeting in being. After a short discussion, the meeting voted to disband; it being understood that there would be no difficulty in calling the Group together again should their new arrangements not work out as well as they anticipated. The Group no longer being in existence, their final decision was to inform higher management that they would no longer be sending a representative to the Works Council.

CONCLUSION

The middle management stratum of superintendents and departmental managers had become displaced persons in the executive system following the extensive organizational changes between 1938 and 1942. The extension of the controlling authority of both the established and the newly introduced functional managerial services had diminished the superintendent's authority over his own sphere of control and had given rise to intense feelings of loss of status. This sense of loss of status and authority was intensified by the setting up of the joint consultative system allowing for direct contact between top management and the workers' representative, and leaving the superintendents feeling squeezed out, for they had not approved of the new set-up from the beginning.

The inclusion of the superintendents in the new consultative machinery by establishing a "democratic" Superintendents Committee, which was given one place on the Works Council, did not obviate the difficulties, and in some ways made matters worse. For the superintendents felt that not only had their authority and status been nibbled away by functional managers and by committees of workers, but that their worst fears were realized in that they were never to be part of a coherently functioning executive chain in which they could meet and discuss their jobs with their own immediate superiors and receive direction from them, but were to be abandoned, and left to their own resources, with one dubious voice in the Works Council. In such a setting of uncertainty about their position in the executive system to which they really belonged, stresses increased within the group itself and rivalry over status became an issue of considerable weight. Interacting with these difficulties within the group was their inability to mobilize a unified point of view to present to higher management, with the result that the experience of these shop leaders was not available for those concerned with more general planning.

It will be noted that one of the main factors making possible the formation of the new Works Managers Meeting was the recognition of the function of joint consultation as essentially a policy-making mechanism to sanction executive authority. This having been clarified, it has become possible for the firm to come to grips with the problem of the exercise of authority in the executive system. This clarification occurred during the period when the Superintendents were busy working through the problems of their own relations, and these factors, added to which was a change in leadership, so combined as to bring about the considerable change recorded.

There is evidence that the new set-up will obviate many of the difficulties experienced by the middle management group. This is not merely because of the organizational change which has taken place, but because the organizational change is just one aspect of more general changes in outlook occurring throughout the managerial system. The top management is accepting full responsibility for all executive work; in his turn the Works Manager becomes responsible for directing the work of the Superintendents, and they for directing the work of their own shops. That such a point seems obvious overlooks the fact that this is not a return to any previous form of organization within the factory. There has been a creative development. The "big stick" or "tough leadership" policy which was maintained until 1940 was replaced by what they refer to as the "soft" or "consultation" policy; this latter has now given way, not to the old "big stick" system, but to a new form of organization for this factory, which can be thought of as a constructive leadership, with real authority derived from consultation, and subject to review by those who are directed. The new Works Managers Meeting provides the central co-ordinating mechanism for this development so far as the production shops are concerned.

CHAPTER EIGHT

THE DIVISIONAL MANAGERS MEETING

Top Management and Executive Leadership

THE personality and methods of work of the chief executive, and the quality of the relations within top management, are among the most important factors affecting the operation of an organization. These features came under review in the course of discussions with the Divisional Managers Meeting, and threw into relief the influence of social factors and group relationships in the making and carrying out of all aspects of factory policy—economic, technical, and social. The account to follow will be concentrated largely on the reporting of work done in sorting out roles and relationships within the top management group in conjunction with their own development of their social and organizational policy. Little reference will be made to the managerial activities in the commercial and technical spheres, except in so far as the opportunity presents itself to show how decisions in these spheres too are influenced by social and psychological considerations. Special emphasis will be given to the demonstration of certain selected problems which may have general interest, and there will be only passing reference to the normally sound day-to-day working relationships among the managerial personnel concerned.

Along with the events which are here portrayed in a highly selected and condensed manner, the Divisional Managers Meeting was effectively engaged in carrying out its day-to-day managerial tasks. It planned and put into action important new technical developments for improving the production of bearings; faced and dealt with the period of redundancy caused by economic forces outside its control; discussed and developed new ideas in joint consultation and personnel management; radically revised and improved methods of budgetary control; and so carried on the many other activities which constitute its normal business as to maintain the firm in a sound financial and trading position. The only constant reminder of the normal soundness of morale in the Divisional Managers Meeting will be the fact that it had the capacity to identify and tackle difficulties of the kind to be described, instead of merely avoiding them by denying their existence.

In short, in order to appreciate the following material, it is necessary to perceive clearly its setting; namely, an ordinary, good management group, running a successful business, and sufficiently secure to be able to seek to improve further its relations and, hence, its managerial efficiency, by carrying on a continuous examination of the nature of its managerial behaviour.

BACKGROUND DATA

The Divisional Managers Meeting was composed of the Managing Director and his immediate subordinates—the Works Director (who is the Deputy Managing Director), the Commercial Director, the Finance Divisional Manager (Company Secretary), the Service Divisional Manager, the Personnel Director, the Medical Officer, and the Technical Manager, with

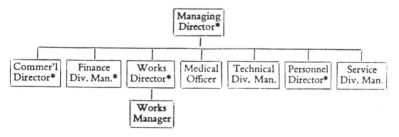

Diagram 15. Divisional Managers Meeting

The position of the meeting in the organizational structure of the factory can be seen by reference to diagram 6 on page 54. * Indicates the members who hold five of the seven positions on the Board of Directors.

the Works Manager in attendance. First organized in 1945 because of the common wish of the Managing Director and his immediate subordinates to meet together as a group, its history can perhaps best be put in the words of the Managing Director in a recent report:

The Divisional Managers Meeting was first brought into being in 1945 at the desire of a number of people, who later became Divisional Managers, to express their points of view in a meeting to each other and to me. I think the emphasis was then upon the desire to speak to me in front of others rather than individually, which had been the practice in the past. This gave a certain form to the meetings which they still retain. The most important part of the meeting was the initial period when each divisional manager reported on matters which he deemed to be of interest to others and made proposals which he felt should get the agreement not only of the Managing Director but of the other members of the meeting. At that stage, the number of issues "put on the agenda" was very limited indeed. The reason for this I think was that these were dealt with as between myself and the divisional manager individually, and further because the number of issues travelling, so to speak, from the bottom upwards was small compared with the much greater volume of instructions, suggestions, etc., travelling from the top downwards. As the process of delegation which arose out of these meetings progressed, the volume of suggestions and proposals travelling from the bottom upwards increased enormously, and the volume in the contrary direction decreased.

The Immediate Complaint

The Research Team's collaboration with the Divisional Managers meeting began in September 1948 out of a request to help them with certain rather indefinite problems. The Personnel Manager, who had been sent to convey the request, explained that in their meeting on 15th September, during a discussion of the sources of managerial authority, the Managing Director created quite a ferment with the suggestion that they might consider questions of final power and authority in the factory with particular reference to the function of joint consultative bodies. This led to such a heated discussion that a number of members of the meeting felt that some sorting out of their own relations, particularly between the Managing Director and the rest, was necessary.

The year before, in similar circumstances, they had held two informal evening meetings, for which the agenda had been a frank discussion of themselves. They thought that not only might further discussion of this kind be useful, but that there might be some merit in having one of the members of the Research Team attend, on the grounds that a more or less independent outsider might help them view their problems more dispassionately. Such collaboration having been mutually agreed, the Project Officer of the Research Team was allocated as consultant, partly because of his contact with this group in the past, before the presently described research project had been undertaken. A description of these previous contacts will give a fuller appreciation of the events which occurred during the period under consideration.

Early Contacts

The Tavistock Institute's first contact with the Divisional Managers Meeting was in April 1947, and was concerned with the wish of the Managing Director to obtain Institute help in dealing with certain rather vaguely defined morale problems in the factory. The Managing Director came to the Institute accompanied by the Medical Officer and the Personnel Manager, the latter two having been deputed by the management meeting to accompany him on these exploratory discussions. The Institute consultants were soon made to realize that the Managing Director was largely alone in his wish for Institute collaboration at that time. Both the Personnel Manager and the Medical Officer, accurately reflecting the feelings of the rest of the management, considered that the Managing Director was trying to rush ahead too quickly. An industrial psychologist had only recently completed two years' work with the firm, and they maintained that they should attempt to consolidate the results of that work before rushing ahead with something new. The Managing Director was impatient with his management colleagues for acting as a brake on his getting ahead with what he considered would be very much to the advantage of the firm. But the Institute was unwilling to undertake to work without the agreement of management, supervision and workers. The Managing Director was in a dilemma which he tried to solve by informing the Personnel Manager and the Medical Officer that the matter was now entirely in their hands; they could either drop the whole idea or, if they wished, could talk it over with management, supervision, and the workers, to see how much interest there might be in having the assistance of the Institute.

As has already been described in Chapter Two, the matter was taken further, the representative bodies did agree, and the Institute carried out an exploratory study in June 1947.

The "Gold Brick" Scheme

The next contact occurred in October 1947 at the beginning of discussions on the "Gold Brick" described in Chapter Two. Because of divided opinion in the meeting on the procedure for getting proposals from the factory for the equitable distribution of the money, help was sought from the Institute. One of the suggestions they were considering was that the employees should have the opportunity to talk over quite openly the relative sizes of increase they should receive; but they were undecided how difficult it would be for employees to discuss their wages with each other. Might they not get some help with their problem, asked the consultant, if they considered it from the point of view of how they themselves arrived at common agreement on increasing their own salaries? There were smiles all round at this suggestion. They did not know each other's salaries, and in fact wished to avoid knowing them, being content, as they put it, to have their interests taken care of by the Managing Director. Thus while management were trying to implement a policy of consultation and to encourage open discussion of all matters, they were in many respects unable to implement these policies among themselves.

The Foundry Trouble

The last of these contacts prior to the current project occurred in December 1947, when a meeting was called by the Managing Director to discuss a report on work which the Institute had carried out in the Foundry.[1] Present at the meeting were the Managing Director, the Personnel Manager, the Works Director, and the Works Manager. It was noticeable that, although the ostensible business of the meeting was to discuss the Foundry report, no representatives of the Foundry were present. The reason soon became evident: the real but largely unrecognized issue before the meeting had to do with the relationships within management.

This unrecognized issue soon obtruded itself and became the focal point of the discussion when the Managing Director stated that he was unsatisfied with the role of the Institute. Another manager had expressed to him dissatisfaction with the Institute's report, and he felt it was only proper that the Institute

[1] This report, and the study which led to it, has already been referred to in Chapter Two.

staff, and not himself, should answer these criticisms. He thought the Institute staff were too diffident and did not stay around the factory long enough. Both the Personnel Manager and the Works Director disagreed, maintaining that the Institute was correct in spending only as much time in the factory as they were asked to do by the people concerned.

The consultant gained the impression that they were talking about their own relations in terms of the Institute, for they did not address each other directly but seemed to speak to each other through him, giving him the feeling of being used as a kind of telephone switchboard. He commented on this and, pointing out that the Foundry report had been left very much on one side, he suggested that the main but unrecognized reason for the meeting was to have him help work out relations at divisional manager level. The reply from the group was that discussions in management meeting never seemed so vehemently emotional when an Institute staff member was present.

The Foundry report was readily forgotten, and replaced by a discussion of differences within the management group. The others criticized the Managing Director for the manner in which he sometimes expressed displeasure, and for his excessive interest in human relations problems at the expense, as they thought, of technological development in the company. They felt they could not influence his behaviour because they were operating what was referred to as an "authoritarian" system in their own management group, in contrast with their attempts to establish a "democratic" system in the works. The Managing Director was father, judge, and executioner of the divisional managers, but although they all complained that he behaved in an inconsistent manner, nevertheless they all contributed to maintaining this state: it seemed preferable to having more to say in their own affairs. In the light of these comments the Managing Director asked, "Suppose I were to abdicate? Surely the same problems would continue?"

The consultant indicated that in some respects the Managing Director had already abdicated, as for example when he had thrown to the Personnel Manager and Medical Officer the responsibility for deciding whether the Institute should be asked to help with some of the factory's developmental problems. This comment came as a shock to them, and started up a critical appraisal of the Managing Director's behaviour at a Divisional

Managers Meeting about two months before, when they had been discussing a report of a management subcommittee on the "Gold Brick". He had disagreed with the report, but finding himself at odds with the rest of the management group, had said "I don't agree, but go ahead and you will see what happens". This behaviour had been taken as his rejecting the group and abdicating, and had given rise to an argument followed by the resignation of the Works Director from the management subcommittee. It was because of this difficulty that the Institute consultant had been called in to the management meeting described above, when the "Gold Brick" plan was discussed.

The Foundry problems, which were to have been the main subject of discussion, were not directly touched upon. This was recognized towards the end of the meeting and the matter referred to the Works Director and the Foundry Manager to deal with. The problems within the management group, however, remained unresolved.

Some Impressions of Top Management

These specially selected events, which occurred before the present research project started, are described in order to give some background and a longer time perspective from which to view the developments which have taken place during the project. There was no further contact between the Institute and Glacier management other than that relating to the iniation of the project described in Chapter One, until their request for assistance in September 1948.

From these early contacts with the Divisional Managers Meeting certain impressions emerged. The Managing Director had been responsible for bringing a great deal of order into the organizational structure of the firm between 1939 and 1945; but he felt unsatisfied with the results of his attempts to raise the level of democratic participation in the running of the firm, and hence was most anxious to press ahead in that particular direction. In doing so, however, he had to some extent lost sight of the uncertainties of his immediate subordinates, and had been unable to get their support to the point where they would tolerate his new ideas, even if they did not fully understand them.

Limited and tentative attempts to resolve this dilemma in the management group had been made from time to time, but with only partial success. As one means of dealing with stalemates arising out of difficulties in arriving at compromise solutions, the Managing Director occasionally resorted to the technique, exemplified above, of sudden "delegation" to his subordinates of full responsibility, and authority to deal in their own way with a problem; he himself having nothing to do with the matter. Such delegation was confused with, or rather rationalized as, democratic behaviour, but was nevertheless experienced by his subordinates as abdication, and interpreted by them as a concealed disciplinary measure.

Thus, although the Managing Director and his colleagues earnestly desired to ensure the success of the consultative policy in the factory, they had more or less overlooked the implications of such a policy for their own behaviour, and had therefore not succeeded in effecting changes in their own relations to each other in harmony with the changes they were trying to implement elsewhere. These difficulties become more comprehensible when they are considered as a hangover from the time in 1938-39 when there had been co-managing directors who had not seen eye to eye.[1] The split in the management which had resulted had never properly been healed, for after the present managing director had achieved full control in 1939, he had to cope with the aftermath of the divided loyalties which had existed. He strengthened the hand of the functional managers in order to compensate for what he perceived as weaknesses in the line management chain, and although this had worked tolerably well through the critical war years, it did not provide a fundamental solution; and the difficulties had to be tackled afresh once the war was over. Much of what follows may be regarded as a serious attempt by top management to implement among themselves the human relations practices they were advocating for their subordinates, and, in so doing, to resolve once and for all the problems from the past which continued to affect them.

SPECIAL EVENING MEETINGS

Arrangements by which some of the management problems could be sorted out were talked over on 26th October, 1948. There was some anticipation that their work as a management group might be improved, and, somehow, the Research Team consultant was to assist in this. Whether he could or could not

[1] See Chapter Two, pp. 27-28.

was an open question in his own mind, and he hastened to let them know that he could give no guarantees that he would be of any help. Since there was little precedent on which to decide the procedure for such discussions, a trial run of three meetings at roughly two to three week intervals was settled on to see of what value they might be. The meetings were to be held on factory premises just after closing time, and not over dinner as had been done the year before, since the consultant, in order to be consistent with the agreed principles of work for the project, wished to maintain as formal a relation as possible with the group.

As a basis for their deliberations, the Managing Director had prepared a document, which became known as the "We-They" document, dealing with relations between superiors ("we") and subordinates ("they") at any level in an organization.

(1) *If "we" decide, "they" reject.*

(2) *If "we" leave them to decide, "they" may take action harmful to the company and therefore to themselves.*

(3) *If "all" try to decide jointly and cannot agree, what then?*

Various possible solutions were then enumerated, dealing with the functions of management, and the representation of the interests of producer, consumer, management, government, and the public, in policy making. How this document was to be discussed, and for whose benefit, was not established.

The first meeting of the trial series took place on 25th November. The Managing Director unwittingly raised a bone of contention when he began by stating, no doubt expecting that all were agreed, that the group was met to discuss how the problems raised in the "we-they" document affected themselves. Most of the divisional managers did not accept this, believing they were to discuss the "we-they" document in general terms, and not in terms of their own problems. Because the consultant was not sufficiently certain of his own role at the time, he did not refer to this difference in outlook; nevertheless it was a matter of the greatest importance, for it reflected a substantial difference in outlook among the members of the group on whether the practical advantages of the illumination and understanding to be gained from consciously studying and knowing their own behaviour were worth the pain and effort to be faced in so doing—a doubt which was reinforced by the deep anxiety

214

and suspicion, common in some degree to all groups, that insight into one's own group behaviour will lead to the release of destructive forces and cause irreparable damage to the group.

Without any declared decision on the goal of the special meetings, they proceeded to discuss general matters arising out of the document, eventually arriving at the big question of whether or not democratic participation was possible. The consultant suggested that they were confusing democratic participation in arriving at policy decisions, with the delegation of executive responsibility and authority to individuals to carry out policy that had been decided. One Divisional Manager thought that this might apply at top management level, but did not think it useful to talk about policy being worked out at lower levels. This gave the consultant the opportunity to mention the possibility of differentiating levels of policy, each level of management working out its own policy within the framework set by the level above, but in consultation with the people concerned; higher level policy, where necessary, being modified in the light of these discussions.

A notion of this kind seemed acceptable, and the Service Divisional Manager said they had not taken seriously enough in his Division this working through towards a concrete policy, and some trouble in getting technical changes had resulted because of it. The consultant added that working through included not only the taking up of technical matters, but also the straightening out of personal relations within policy-making groups.

They gradually switched back to some of their own problems, including their difficulties in arriving at decisions satisfactory to them all, and the cluttering up of their business meetings with matters of detail interesting to two or three only of those present. As though wishing to close the first special meeting with some positive action, they decided they would all co-operate to eliminate these detailed matters so they could deal exclusively with more general policy questions.

Purposes of the Divisional Management Meeting
The thorny question of the goal of the meetings was for the time being set aside—and a positive, constructive purpose served as well—by concentrating in the second session, on 20th December, upon the question of the structure of their meeting. Was it the Managing Director's executive meeting with his subordinates,

or were they a true management committee taking decisions as a group? This question went back to the time in 1939 when the present Managing Director had taken over and not all of the senior management personnel favoured the new policies introduced. They had maintained a wait-and-see attitude for a number of years until their doubts were resolved, these doubts in the meantime producing stresses between them and the divisional managers whose task it was to implement the new policies. As one step towards reducing doubts and apprehensions, the Managing Director had guaranteed that final authority would continue to be vested in his own role, and in himself as long as he continued to occupy it, and not in the Divisional Managers Meeting as a group.

The situation was made more complex by the fact that five out of the nine members of the Divisional Managers Meeting were also members of the Board of Directors. The consultant indicated to them how these influences together had the effect of allowing the Divisional Managers Meeting to function sometimes as a concealed Board of Directors concerned with the operation of the company, sometimes as an executive management committee taking decisions affecting the London factory, and sometimes as a meeting of divisional managers for the discussion of mutual problems with the Managing Director but not taking decisions at all. Did this not lead to confusion as to the function not only of the meeting, but of the individual members, which in turn would tend towards inconsistency in behaviour?

The point about role clarification did not get far, however, and towards the end of the meeting one member suggested, and the group concurred, that the consultant should attend their regular weekly meetings, as this would afford him first hand contact with their difficulties. The consultant consented to give this a trial, and although technically it represented a new departure at that time, eventually much of the activity on the project took the form of working with groups as they carried on their day-to-day business.

ATTITUDES TOWARDS AUTHORITY

The third and last meeting of the agreed trial run took place on 5th January, 1949. It opened with informal chat about the time of the next meeting, the Managing Director being asked

216

to arrange it, although it had not finally been settled that any further meetings would be held. The consultant commented that this chatting about times, while innocuous enough in one sense, could also be taken as a concealed expression of antagonism towards these meetings; for were they not in effect saying to the Managing Director, "If you want these meetings, then you call them". The Divisional Managers would have none of this until it was pointed out by the Managing Director that their normal procedure was for the whole group together to arrange times, and that for some reason or other they had now departed from this procedure . The others then decided it was correct for the Managing Director to call the meetings, for after all it was surely their purpose to give him help with a problem—help which they were all very willing to give. The consultant took up this point, referring to its inconsistency with the problems that had in fact been coming up, such as the feeling that management meetings were unsatisfactory, and that the Managing Director imposed decisions on people.

The Service Divisional Manager then broke in, querying whether the group considered it was getting anywhere. But the Managing Director thought such a question would be better taken from the point of view of whether the meeting should continue. The Service Divisional Manager replied, "Fine, I will defer to the Managing Director's suggestion," but others immediately said they preferred to discuss the question as the Service Divisional Manager had put it. The consultant interpreted this preference for the Divisional Manager's view as partly a reaction to the Managing Director's suggestion, arising out of the feeling that he was imposing, or trying to impose, his own point of view on the group or on individual members of the group. He added that this reaction to a feeling of being dominated seemed mixed with its exact opposite: a desire to have an autocratic boss who would tell them precisely what to do and tolerate no indecision.

The Service Divisional Manager changed the subject, saying he would like to make a suggestion which he hoped would help the Managing Director with his problem. He then went on to explain that the Works Committee members of the Works Council probably resented the fact that the management discussed separately among themselves all items that appeared on Works Council agendas, and came to the meetings with their minds already made up, and further that there might be a general lesson for all of them in this. This idea was immediately picked up and talked over by the others. The way it was seized led the consultant to believe that they were using the Works Council as a transitional situation by means of which they could both continue to evade talking about themselves, and at the same time take one step closer towards a more direct consideration of their own affairs. If it were true, he said, that they were using the Council in this manner as a safe way of talking about themselves, then it would follow that what they were really talking about was their feeling that the Managing Director came to management meetings with his mind made up, and then attempted to impose his ideas on people.

There is little doubt that their talk about the Works Council represented a growing sense of freedom in the group to move towards taking their own relations directly into account. But their anxiety was nowhere near being sufficiently diminished for them to face at that moment the implication of their talk which the consultant had pointed out. Hence the idea that they were talking about the Works Council as a safe way of talking about themselves was rejected by nearly everyone; and the Medical Officer, taking up this theme, denied that there were any bad feelings towards the Managing Director. He thought it would be more correct to say there was occasional mild resentment over the way he handed out work. For example, someone had confided to him that when the Managing Director came up with a new idea, that person would wait for six months to see whether he remembered it, and, if not, would just file the notion away without saying anything. But this was a matter of amusement and not of hostility or real resentment that led to enmity. The Works Manager agreed with the Medical Officer. They were not hostile to the Managing Director because they were dependent on him, but they did feel that at times he was inclined to be too censorious, and too impulsively and sharply critical of them. A number of people concurred in saying that this expressed their own point of view.

In part the Medical Officer was correct. It was not just that the Managing Director was a person about whom his subordinates had such mixed and disturbing feelings: he was also a figure of authority, conjuring up all past experiences with authority, good and bad. It was because there was so much displacement on to the person of the Managing Director of private and distressing feelings that had to do not with him alone, that there was also worry and uncertainty in the group about tackling the problem of authority; but, since it was time for the meeting to break up, there was no chance to pursue the matter then, and another meeting was arranged for February 2nd.

At the beginning of this next meeting, the Managing Director was asked if he could say what he was getting out of their discussions. He replied that, although he could not give any concrete reasons, he had a strong impression that they were getting somewhere. The Commercial Director and the Finance Divisional Manager both agreed, referring to their experiences in a group they had attended at the Tavistock Institute the year before, which never at any time had dealt with anything concretely in the sense of reaching a practical conclusion, but nevertheless at the end all had felt the experience had been worthwhile.

The discussion of whether the meetings were worthwhile was a veiled reference to the role of the consultant, and represented a shifting onto him of those attitudes towards authority which had been brought out in the previous meeting as displaced on to the Managing Director. The consultant had become a transitional figure upon whom the group could test out its attitudes; he was now the authority figure imposing his views on them, and by his remarks generally making the meetings less enjoyable than they otherwise might have been. If he could stand up to criticism and attack, then he would have served a real purpose, for it would enable them to feel less afraid of the consequences of their criticisms and so feel much safer in tackling aspects of their relationships which, up to then, they had not dared touch upon.

The consultant mentioned these thoughts, adding that it appeared that people wished to criticize him, but did not feel they could do so directly. This was quickly denied by the Service Divisional Manager, who said that he criticized the consultant whenever he disagreed, and by the Commercial Manager, who said that if he felt critical, he was not conscious of it. However, following similar denials by other members of any consciously critical feelings, a considerable depression settled on the discussion, which up till then had been animated.

Taking this gloom as evidence of the group's resistance against allowing themselves to recognize the discomfort they felt with him, the consultant said he thought it was an immediate example of their prevailing attitude; for if he were accepted as the authority at the moment, they either feared, or felt it would be impolite, to criticize him directly, and the result was this embarrassed, depressive silence.   Although this was again denied, the consultant sensed that something akin to what he was pointing out was going on, and so he went on to comment that the depressed, silent behaviour might not be the result only of embarrassment about expressing critical views, but might also be one very effective way of expressing their feeling that the special meetings were really not very useful.   Indeed, the group seemed to him at the moment like one in which a person in authority had made a "suggestion" to his subordinates and then asked if they had any comments, and everybody had thereafter stood around in rather dull silence, but had had plenty to say afterwards which they would not say at the time.

At this point the time to finish had come, and they turned to consider whether they were to hold any further meetings. The Medical Officer said he could see a point to the meetings going on if they were to discuss more concrete issues. He would have liked to see a discussion of the general principles of group behaviour so that from these principles they could learn how to treat their subordinates. In his estimation the management group had few internal problems, and he therefore was very much averse to "dangerous introspection", because it would lead to the creation of difficulties where none now existed.

Reversing his previously expressed positive attitude towards his experience in the group at the Institute, the Commercial Director also stated he was most unhappy with these intangibles; only if they could discuss something concrete could he see any point in continuing.   When other members concurred, the consultant, who had to leave for another meeting, once again emphasized that, under the Research Team's conditions of work, he could not take part with management in discussions of other groups in the factory, but could only take the role of discussing with them a topic such as group behaviour in terms of their own relations with each other.   He then left, feeling that they could hardly wait for him to be out of the room so that they could settle down and talk about what had been going on.

Work on the Problem of Authority

Difficult questions of attitudes towards authority had been touched upon in the group and anxiety had been left at a high level because of the consultant's failure to get at some of the roots of these fears of expressing feelings about authority. Nevertheless, his method of approach, which included not only accepting but actually ferreting out criticisms and hostility displaced onto himself, combined with his use of such material not to attack them in return but to try to help them understand themselves, meant that he was both a useful and reassuring transitional figure for them. A further discussion was arranged for 2nd March, but, partly to get away from an examination of the nature of their own relationships, they decided to confine their attention to a point-by-point consideration of the "we-they" document. The meeting opened formally with the Medical Officer beginning to read through the document, but pausing after the first paragraph to enquire if they wanted him to continue, or to stop and allow them to discuss it thus far. Leaving this question aside, the Service Divisional Manager returned to the older problem of whether they were going to discuss the document in general terms or in terms of the Divisional Managers Meeting itself. The Medical Officer replied with some impatience, saying that they had surely agreed already that this was a dilemma only for the Managing Director, and that the rest of them had few problems in relation to each other, with the possible exception of the fact that the Managing Director believed they all had problems.

The seriousness of the group in relation to its task asserted itself and, taking the lead given by the Service Division Manager, the Works Director declared that he was not now sure whether it was true that the rest of them had no difficulties. There were occasions on which he went out of Divisional Managers Meetings with things on his chest that he would have liked to talk about, and which he soon sounded off about when he got out. There were affirmative nods from everybody. The Managing Director said he thought that other people often rejected his proposals largely for no other reason than that it was he who made them, but that they did not feel free to state, except in this indirect way, their resentment that he could impose his ideas upon them. The others replied that the trouble was not that he made proposals, but that they felt he had already made up his mind; and after

221

much debate a consensus of opinion emerged that difficulties arose not so much from anyone making proposals, but from people appearing unwilling to change their minds no matter what happened.

It suddenly occurred to the consultant, and he pointed it out, that there had been a very radical change in the quality of the discussion, for, having begun by agreeing to limit themselves to the "we-they" document, they were now reviewing possible sources of hostility and rejection in their own group without apparently being aware that they had denied the very existence of such feelings just one month before. Furthermore, they were talking about difficulties as existing in the group as a whole, which up till then they had only been willing to admit as the problems of the Managing Director alone.

A number of the divisional managers had also become aware of this change. The Service Divisional Manager thought that in his case it might well be because of the greater attention he had been paying to group relations problems, and the Commercial Director suggested that perhaps the meetings themselves had had some effect in helping them to see more clearly in their day-to-day life how their behaviour affected others in their own divisions, an idea which seemed reasonable to some of them who thought they were more sensitive to the attitudes of their subordinates than they had been before.

Acceptance of the fact that they all shared and experienced the same problems as those of the Managing Director, both as his subordinates and as leaders in their own right, allowed a more detailed discussion of the Managing Director's behaviour to take place than had hitherto occurred. It was suggested that when his proposals were sometimes rejected, it was not simply because he made them, but because on those occasions the others suspected he might have a particular bias; as, for example, a bias towards the human relations rather than the technical side of things in the factory. But even this was not the full explanation, for at least one other factor which entered in was the rate at which he often tried to force the pace when he wanted things done; the faster he tried to push, the more he was resisted, particularly if the others had not yet made up their minds whether they thought the plan a good one, and there seemed time for further discussion.

Although the meeting had perhaps thrown up more new problems than it had resolved, the growth of security sufficient to make possible the open consideration of some of the feelings previously unexpressed, or only indirectly expressed, produced a sense of satisfaction. The Commercial Director summed up the general mood when he remarked that for the first time he felt the meeting had been really worthwhile. Nevertheless there remained so far untouched a completely different but related set of difficulties, namely, differences between the divisional managers themselves. Some of the satisfaction which was now being expressed was undoubtedly due to the fact that their relationships among themselves had been successfully avoided by the exclusive concentration upon the Managing Director. But the fact they could now talk about the Managing Director meant that there was one less barrier protecting them from these other issues, which they were very soon to encounter.

Recoil from Rivalry and Competition

During this time the consultant had been attending the weekly Divisional Managers Meeting. It was the period when the redundancy crisis had hit the factory and, apart from preoccupation with the difficulties of the general economic situation, an important task of management had been to ensure that with the lessening of production, and the release of production workers, there should be an equivalent reduction in overheads, as represented in the activities of the non-production departments. This had led to weekly discussions in the Divisional Managers Meeting of the cutting down of indirect staff, with each divisional manager responsible for effecting this in his own division. The point had been reached, however, where the divisional managers felt themselves unable to achieve any further reduction except by radical measures which might seriously impair efficiency. They recommended that the Managing Director should take over the responsibility for deciding whether some activities carried on in the factory might not be cut out completely. He replied that he had been unable to discover any reasonable rearrangement of divisional responsibilities which would improve the situation, and hence wished the divisional managers themselves to continue to continue to carry the responsibility for staff reduction.

The consultant commented that their manner of approach to the redundancy problem had much the same quality as some of the discussions during the special evening meetings. They seemed always to deal with only two alternatives—it was either the Managing Director or the individual divisional manager who was to take responsibility for decisions. The possibility that they might all together discuss the relative merits of certain functions in one division versus certain functions in another did not seem to occur to them. The Works Director and the Commercial Director, agreeing that this was correct, went on to make the significant observation that in this they were very much the product of the executive atmosphere in which they had grown up. The notion of people with equal status discussing with each other the interdependent problems of their respective spheres of influence was outside the culture to which they were accustomed.

The chief mechanism for dealing with any stresses had for a long while been either to avoid them by not recognizing them, or to get rid of them by routing them through the Managing Director. But some of these unrecognized stresses were now beginning to demand attention; a process which was further stimulated when the matter of the relations between the divisional managers themselves came up again at the next special meeting on 30th March, just four weeks after the previous one.

They returned to the "we-they" document, with particular reference to whether it was true that if a person in authority made a suggestion, it would be turned down by his subordinates merely because the suggestion had come from higher up. The consultant observed that it had appeared to him from management meetings that there was a tendency for them all to reject each other's suggestions. He had been struck by the limited extent to which the divisional managers took up suggestions made by each other; it usually being left for the Managing Director to comment first. Did this not indicate some lack of patience with each other's suggestions; and was not the Managing Director being used to mediate between them, and turned into a "perfect person" in accordance with their wish that he should solve their problems in this way? If such was the case, it would suggest that there was possibly some element of competition among the divisional managers. Such competitive feelings could conveniently be concealed by maintaining the code that it was inappropriate for executives at the same level to sort out differences with each other.

The mention of competition led to a vehement denial from most of those present, except the Managing Director and the Personnel Manager, who felt that there must inevitably be some competition, though perhaps people were not aware of it. One of the divisional managers replied, and in this probably expressed the most general feeling in the group: "It may be that it is unconscious, but I certainly have absolutely no awareness of jealousy, competition, or rivalry with any of the other divisional managers, and in fact the whole suggestion just leaves me cold."

The suggestion about rivalry and competition within the group had not been clearly or adequately put by the consultant. He felt that something of this kind was operating but did not elaborate the point with sufficiently concise illustrations from their behaviour at the moment, and thus failed to deal with the emotion which such an interpretation inevitably aroused. It may well be that the amount of feeling expressed in the rejection of the idea was itself both an indication of the correctness of the comment and of the clumsy and incomplete way that it was put. However, the time for finishing had arrived.

Because two of them had to go off to meet some customers, the meeting should have ended promptly. In spite of this they continued for a short while longer, as though reluctant to drop the matter, the Managing Director asking whether the divisional managers had not quite commonly observed competition among their own subordinates. They all admitted that such behaviour occurred frequently enough, but they had noticed that their subordinates were reluctant to face these issues among themselves. In most cases of differences of opinion, they preferred to state their cases privately to the divisional manager, and let him decide. There was no opportunity for the consultant to take up the fact that this behaviour attributed to their subordinates might equally well apply to the divisional managers themselves. The meeting had gone on much longer than planned, and broke up without any date for a subsequent meeting being set.

Conclusion of the Special Evening Meetings

From this point onwards, no further special evening meetings were arranged for nearly a year, and on looking back, there inevitably arises the question why they just petered out at that time in the way they did. The impression is left that one factor which contributed heavily was the consultant's failure, partly due to his uncertain role, to take up more thoroughly, at the very beginning, the nature of the meetings and of his own role at them. This meant that what was possibly one of the main problems, although touched upon, was nevertheless not effectively worked at; namely, the differences in attitude of different members of the meeting towards discussing matters which had directly to do with the working relationships within the group itself. These attitudes varied from the fear that dangerous introspection was being indulged in to the point of view that there was no other way to consider problems of executive relations except by considering their own.

How the group reacted to the possibility of gaining insight into their own relationships can be seen in the course of the special meetings. There was at first complete denial that anyone but the Managing Director had any problems whatever; it being argued that the others only wished to help him with his difficulties. That this co-operative attitude towards the Managing Director also concealed more negative aspects was only allowed to emerge after the group successively used the Works Council and the consultant as transitional objects upon which to test out their negative and critical attitudes, discovering in the process that such attitudes could not only be tolerated but also looked at for the constructive purpose of throwing light upon how they came into being. Hostility towards persons in authority was not just a case of personal dislike of the individual, but a matter of complex attitudes left over from past experiences which were touched off and displaced on to the nearest person in charge. Recognition of this fact of displacement made possible a movement forward in the group towards a more constructive consideration of the type of leadership given by the Managing Director.

This move forward then received a set-back when it led right on to another problem which had been lingering about in the discussions from the very beginning: that of the relationships between the divisional managers themselves. The discomfort aroused by this particular problem serves to explain a number of the features of the evening meetings. For example, the "we-they" problem could not be tackled because it was incompletely formulated; a full statement required the inclusion of the "they-they" and the "me-you" problem as well. That is to say, the relationship between a person in authority and his subordinates could not be tackled in isolation from the relationships among the subordinates as a group, as well as the real personal relations between individuals. That these problems were so touchy is understandable when it is realized, as the group eventually discovered, that they were tied up with the sorting out of policy on the power and authority of the functional and the line managers within the group, as well as with evaluating each other's capacities in leadership.

The argument as to whether the Managing Director was trying to advance too quickly on the human relations front at the expense of technical development was also used as a mask preventing a too intimate contact with the deeper-lying questions of authority and rivalry in the group. In fact an extremely active technical developmental programme was being carried out, and examination of the Managing Director's deployment of his own time left no doubt that he was the active leader of the firms's technical work. If, however the convenient view could be upheld that there was a split between the Managing Director and his divisional managers on social versus technical policy, then certain advantages would accrue to everyone. The Managing Director on his part could say that, since his divisional managers were so backward, he had no alternative but to overlook the question of his own relations with them, and to rush ahead finding support wherever else in the factory he could. The divisional managers in turn could say that their only difficulties had to do with the Managing Director's impatience. By this type of unwitting collusion they could all to some extent avoid having to deal with the human relations questions which really affected their managerial and administrative work; namely the exceedingly difficult problems of superior-subordinate as well as subordinate-subordinate relationships; while at the same time allowing some progress in the firm, because none of them would take a strong stand against the Managing Director's new ideas.

CLARIFICATION OF ROLES AND PROCEDURES

Following the discontinuance of the special evening meetings, the main focus of work with the Divisional Managers Meeting moved over into their regular weekly business meetings, which the consultant had been attending since he was first invited early in January. In these meetings he took occasion to comment from time to time on how the group was discussing its problems, with particular reference to the manner in which the relations among members of the group affected their work.

On 17th April the question of the cutting down of services and restructuring of divisions in order to cope with the problem of overheads due to redundancy came up again. The divisional managers argued once more that the Managing Director should take such decisions, but he reiterated that, although he had considered the problem a great deal, he had not been able to see any reallocation of responsibilities which would help. He still believed the detailed readjustments he had proposed were sufficient if properly implemented, and asked why, if they were dissatisfied, the divisional managers could not come forward with any suggestions. His question was answered by two of them to whom it seemed that the rivalry within their own group which they had seen in their last special meeting was with them again. Perhaps the truth of the matter was that inter-divisional reorganization was difficult because there were too many personality problems in the way. Other members of the group, however, shook their heads in disagreement.

An Experiment in Chairmanship

A week later the Managing Director brought forward a new suggestion that someone other than himself should be chairman, on the ground that such a move would allow freer discussion from everyone, including himself. He had not perceived, however, that he was indirectly raising the more fundamental question of the nature of the Divisional Managers Meeting. In the opinion of the consultant, if the meeting was just an informal getting together of the Managing Director and his divisional managers, it did not seem to matter much who was in the chair. But if it was an executive meeting, was it not automatic that it should be the chief executive, since he was the person responsible for the decisions taken? Without resolving these questions they decided to see what it would feel like with the Managing Director as an ordinary member of the meeting. The Commercial Director was elected chairman for a trial period of one month.

Tied up in the trial of a new chairman was the issue of the nature of the relation between the roles of the Managing Director and the divisional managers, in particular the balance between authority and delegation of responsibility, as well as the problem of rivalry. In one sense it could be said to be a highly skilled intuitive move by the Managing Director to see that the special evening meetings were continued. In order to help get the thorny rivalry problem settled he symbolically "joined the group", and by so doing he partly shared the problem, and partly also unwittingly re-created something like the special evening meeting situation in which he had not been the chairman. As well as taking his place with the subordinates, he also offered fuller leadership by allowing himself greater emotional contact with his subordinates and their difficulties. Some impression of this process of getting closer can be gained from a document he circulated on May 3rd, in which he set out a list of priorities for each division. The preamble of this document read as follows:

We are going through a rather difficult time at the moment, and I think it may be helpful to Divisional Managers for me to make a statement of the primary issues for which I am expecting you each individually to take responsibility. Before setting these out under Divisional Headings, I would like to say a word or two about my own position vis-à-vis Divisional Managers.

At a time of relative crisis like this, I think it is usual for the Chief Executive to take over the reins, so to speak, to a much greater degree than is usual in more normal times, and begin issuing a series of instructions to meet the current situation. It has been a great temptation to me to behave in this manner, but I felt all along that it would not be in the best interests of the firm for me to do this, and that if I did so, it would put back the clock quite a bit. By this I refer to the growing devolution of responsibility which we have all been a party to over the last three or four years, which I feel is gradually producing a situation where in response to an emergency a growing number of people automatically "go into action".

It is partially this feeling that actuated the suggestion I made to you at the Divisional Managers Meeting on 26th April, about chairmanship. My own role is changing and I feel that if I can co-ordinate the activities which arise out of the responsibilities devolved on to you, without having to take too active a part in the discharge of detail, then I will be free to do more in thinking about the future and I will be able to be "on tap" continuously during a period when it is very necessary that I am freely available. To this end I am cutting outside activities as much as I possibly can at the moment and I propose to spend more time in the Works in future than I have done over the past two or three years, being freely available for consultation at your demand but largely leaving the discharge of major responsibilities to you with an increasing degree of non-interference, except

229

*when my counsel is sought. You may, of course, like to challenge
this viewpoint and discuss it, which you are very free to do at a
management meeting.*

A further contribution to the analysis of the role of the Managing Director was made in a special report presented by the consultant to the Divisional Managers Meeting in June. This report singled out what at the time appeared to him to be one of the bottle-necks in the management group; namely, the responsibility unwittingly taken by the Managing Director for personnel matters, because he felt the Personnel Manager was not carrying out the whole of his job. This seemed to have led to his adopting part of the role of the Personnel Manager in addition to his other heavy tasks, using the actual Personnel Manager as a kind of assistant, and as a result he was less available for developmental work in connection with some other divisions and their activities.

To sort out this mixing of roles, a discussion was arranged between the Managing Director, the Personnel Manager, and the consultant. Some clarification of the executive and professional roles of the Personnel Manager led on to further elucidation of the role of the Managing Director himself, particularly with regard to his behaviour at Works Council meetings. This last was a position of particular difficulty for him, because he strongly desired—and thought he ought to be able—to speak for himself as an individual in the Council setting, for, after all, was he not only an ordinary member just like everyone else? The Personnel Manager, however, was quick to inform him how often his personal comments disturbed people because they could not dissociate him from his role as chief executive. In other words, whether he liked it or not, anything he said would be taken as coming from the chief executive, and not from himself as a person. In effect only two courses of action seemed possible for a managing director at Works Council; either he could take it upon himself officially to commit management to a line of action, or to state that he would refer the matter back to management. Personal comments, indeed, seemed out of order.

One other matter which came out into the open was the considerable preoccupation of the Personnel Manager at that time with visitors to the factory, an involvement which drew him away from his other duties, and which took up the time of the

Managing Director as well. Were they not, asked the consultant, at least equally concerned with spreading information about their social developments to other firms as with taking up difficulties in group relations within their own factory? If so, could this not reasonably be considered in part as a convenient and effective means of escape from a difficult situation; since it was easier by far to talk to visitors about good human relations than to tackle the difficult social problems facing them in the factory?

The Personnel Manager confessed that he was often embarrassed when explaining to visitors the methods they were using. He knew very well that the structure he was describing, while it looked nice on the surface, did not work all that well if you really went into it. The consultant interjected that they seemed to have become caught up in an unwitting competition with other firms which prided themselves on their social policies, and to be striving to keep ahead of the others by introducing all the latest ideas, and unconsciously trying to give as good a picture of themselves as possible. In this way they were able to get rid of some of their competition with each other inside the firm.

Some Changes in Procedure

During this period the Managing Director had produced another document in which he tried to clarify top management procedure, and of which the following are the introductory paragraphs:

> *I think most people are worried at the moment because the time taken in discussion of affairs at top management level is tending to increase, and the range of problems that appear to warrant discussion at this level continues to widen. It may be that people are feeling less effective than they were, and I am prone to this feeling myself. It is perhaps as well to get the matter into perspective.*
>
> *I have been Joint Chief Executive and Chief Executive for eleven years, and looking at it from this rather historical perspective gives me a great deal of confidence, as against the lack of confidence I feel looking purely at the current situation. Looking back it seems to me that in the past there was never anything going on that I had not had a finger in starting, and my chief feeling in the past was one of anxiety that more was not being done in a wide variety of directions. A decisive change has taken place since then, because my present*

anxiety is that there is a great deal being done that I ought somehow to be co-ordinating and do not find the time to attend to. Along with this I have got a feeling of pronounced satisfaction that there are all sorts of initiatives taking place that I know nothing about until they are fait accompli, or very nearly so. This seems to me to be a very healthy position, even if occasionally it can give rise to a few mistakes. In other words, we have changed over from a position of some apathy, considerable lack of activity, coupled with an ease of control and a lack to progress, to one where there is a very great deal of activity, great difficulty in co-ordination, but a vast volume of spontaneous thinking and action going on. Looking at the big picture, I think it is very encouraging. . . .

Arising out of the memorandum from the Managing Director, a number of procedural changes were made, including the introduction of a secretary into the meetings to keep minutes— a practice which had been discontinued three years before—and setting up new methods of routing business through the Managing Director and for delegating responsibilities. These changes were parts of the general move towards increasing efficiency (although the Service Divisional Manager expressed strong concern that too much formalization was being introduced), and towards removing such anomalies as their inability to refer with any precision to decisions they had taken when they were not keeping minutes. Their rather informal methods had stood in the way of their building a framework of recorded policy, and hence they had been forced to depend largely on what people could remember.

Along with these alterations, they recognized that although the Commercial Director had done an excellent job as chairman, nevertheless it did seem more realistic for the Managing Director to take the chair once again. Merely changing the chairman was unlikely to help much in building the most effective pattern of relations between the Managing Director and his team. With the Managing Director back in the chair, they could get back to tackling their top level leadership problems more directly. The experiment in chairmanship was thus a complete success in the sense that, by allowing direct experience of one other form of organization, it made possible a return to the previous form, but at a new level of insight into the nature of the executive process.

The Role of Managing Director as General Manager

The clarification of the role of the Managing Director and indirectly of his subordinates was carried a step further in a discussion between him and the consultant on 2nd August. On the organizational chart he appeared only as Managing Director of the total concern:

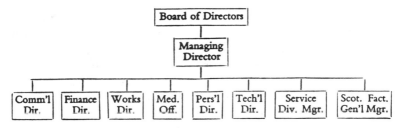

Concealed by this chart, however was an additional role which had never been made explicit, namely, that he was also the General Manager of the London Factories, as shown in the next diagram:

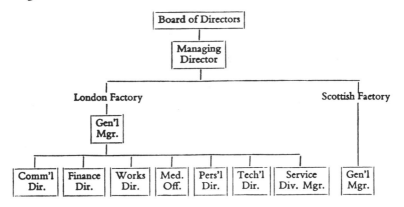

In other words he had not only the job of co-ordinating the work of both the London and Scottish Factories, but also that of running the London factories. That this was more than an academic point became clear once it was realized that it meant that the organization had only a part time managing director, and the equivalent held true for the London Factories with only a part time general manager. This meant, in the case of the London factory, that very considerable responsibilities would have to be delegated to the Divisional Managers. But was this being done?

This last question was partly answered by some strikingly frank talk a few days later. The General Manager (as he shall be designated from now on) maintained that he had been fully delegating responsibility, except for new social and consultative developments for which he had retained responsibility for want of confidence in the others. His lack of confidence was immediately challenged as unwarranted; for whatever might have been the case in the past, they were now sufficiently interested and sufficiently skilled in consultative practice to allow for greater delegation of responsibility. To this view the General Manager acceded, an action made possible because their organizational problem had been perceived not just in terms of a lifeless rearrangement of the organizational chart, but in terms of the implications of changes for the actual behaviour of real people in relation to each other. This perception had led on from the clearer recognition of the General Manager's role made possible by the experiment in chairmanship.

ELUCIDATION OF MANAGEMENT POLICY

The work thus far had dealt largely with the clarification of roles and relations within the Divisional Managers Meeting. Concomitant with some progress in these directions, certain difficulties associated with not having a conscious and recorded management policy moved into the forefront, along with other difficulties in two-way communications within top management as well as between top management and the bodies for which they were responsible. These moves partly arose from the discussion of similar matters at the Works Council week-end conference early in September, which gave additional impetus to the increasing formalization of certain basic aspects of their work which the top management group already had under way.

To begin the long process of getting out a written statement of the policy being operated by management, the General Manager and the Personnel Director (as he now was, through recently having been elected to the Board) were to prepare a document setting out as much as they could remember of the main decisions which had been taken from the beginning of the Divisional Managers Meetings. At the same time the Medical Officer had pointed out that they really had no set of principles which would allow them to decide which types of information should be communicated to the factory through the consultative channels, which through the executive channels, and which through both at once. In order to build up case law by which to sort out this thorny problem, they decided to consider every item they dealt with in order to see if it should be communicated, and if so, through which channel, a procedure which has been used since.

A Clash over the Trade Unions

On 22nd November, as though to punctuate the need for a more concretely recorded policy, an event occurred which drew attention to lack of agreement on certain matters of policy about which there had been apparent agreement, at least verbally, until that time. The General Manager brought to the Divisional Managers Meeting a document (reproduced in full on page 143) which had been drawn up by the Glacier delegation to an inter-firm conference, and in which was proposed greater co-operation and full participation by the unions in setting policy for the factory through the joint consultative machinery. He took it for granted that the document would readily prove acceptable to all. But the opposite was the case. Most Divisional Managers felt that he had gone far beyond what had previously been their policy; the main point of difference centring around attitudes towards the factory becoming a closed shop, although the document itself did not specifically mention the closed shop issue.

Somewhat taken aback, the General Manager asked whether he was not correct in assuming that they had all been in accord on three principles with regard to union organization: that people should join their unions; that management, however, should take no part in coercing people to join—this being a matter for the workers themselves; and that if 75 per cent. of the workers were in favour of such a procedure, all new entrants should be members of, or else join, a trade union.

The first two points presented no difficulty, but there were marked differences in interpretation of the implications of the last, the attitude of most divisional managers being that while they would unhesitatingly accept the closed shop principle if this was desired by the workers, this did not necessarily mean that they themselves preferred it as a method of organization. It was in this last regard that they differed from the General

Manager and from the tone of the document which they thought contained an implicit approval of the closed shop principle. To some extent the open emergence of anxieties about the closed shop could be seen as related to the greater awareness and acceptance of differences in the top management group; but if these differences were to be accepted, the General Manager now appeared as the instigator of a new threat in apparently encouraging the unity of the workers.

For the first time for over a year the General Manager displayed open annoyance with the others, saying they had let him down. He believed that the document he had helped to draw up was quite consistent with the attitudes of his team, but now he found some of them expressing a point of view he felt sure he had not heard from them before. It was extremely difficult to carry on if they were going to let him down like this, and not express their real feelings.

The matter rested in this unsatisfactory state until the following week, when the Works Director and the Service Divisional Manager sought to explain the apparent divergence of opinion as being due to the fact that two years had gone by since they first touched upon this problem. It was to be expected that people might gradually change their attitude without recognizing the change until a specific problem arose such as at the present time. The General Manager interjected that he could not be responsible for people changing their minds without letting him know, and repeated that he felt he had been let down.

The consultant took up this last remark as containing an issue of some importance: on whose shoulders did the responsibility lie for ensuring that there was operational and not just verbal agreement on policy? Or, to state it more generally in an executive group, did final responsibility for ensuring policy agreement rest with the leader, the subordinates, or both? If the General Manager's comments were accepted, then following down the line it would be the divisional manager's subordinates and not themselves who would be responsible for ensuring policy agreement at top level in each division. If, on the other hand, it were taken as the responsibility of the leader to ensure that he had the support of his subordinates for his policies, then in turn the divisional managers would be responsible for their own subordinates; and in the present meeting the General Manager could have no one to criticize but himself for any breakdown in policy agreement because of failure of communications within the group.

The group had little difficulty, verbally at least, in choosing to place the responsibility squarely on the shoulders of the leader. It was up to the General Manager in the case of the Divisional Managers Meeting, and up to each divisional manager in the case of his own subordinates, to get operational and not verbal agreement in his working group; the difference being nicely defined by the Works Director as the difference between getting mere acquiescence and agreement which influenced behaviour. They were able to take such a decision because they had by now resolved to some extent their own differences, or at least were less anxious about them. This in turn had allowed them to adopt a far more secure outlook towards the leader of the group, the General Manager, no longer expecting either perfection or self-denial, but capable of accepting his new ideas with less fear and with a willingness to accept the inevitability of occasional mistakes, and to tolerate them when they occurred.

A Company Policy Document

The discovery by the top management group that there could be unrecognized disagreement among them on matters of serious policy led them to hurry on as quickly as possible to develop their written policy statement. Months of hard work between December 1949 and June 1950 led to the drawing up of a comprehensive statement of general policy covering the company as a whole, which has been submitted to the Board of Directors for its consideration, and which comprises sections on: definitions of policy; the purpose of the company; financial operational policy; factory councils and joint consultation; conditions governing individual rights of appeal; an outline of the structure of management; principles governing the manning and operation of the company structure; and conditions of service for members of the company. This document is reproduced in Appendix One. In addition, a series of Management Standing Orders is being built up, consisting of instructions and regulations governing the implementation of all aspects of policy in the London factory. Sets of these Standing Orders, modified as necessary, are kept at strategic points in the factory, where they are available for ready reference to anyone who requires access to them. By this means considerable order and uniformity of procedure are being introduced into the executive work of the factory, where it is a case of implementing ordinary regulations on which there are agreed principles for action.

TWO-WAY COMMUNICATIONS WITHIN THE EXECUTIVE CHAIN

Recognizing that the prevention of unnoticed disagreement in the group demanded that they should be able to communicate effectively with each other, the Divisional Managers Meeting turned its attention to the question of communications among themselves and within the executive chain, initiating the consideration of this problem by minuting on 22nd November, 1949, that "Management Meeting communications with other ranks of management are not satisfactory". To assist in sorting out some of the factors affecting communications, the consultant was to see each of them individually and then to make a report to the group, using the findings from these interviews interpreted in the light of his first-hand experience at their meetings. He would concentrate largely on an analysis of their own assessment of the quality of their relations with each other, because it was in this area that the sources of blockages in communication were most likely to be found.

One of the divisional managers aptly expressed what were the most general views of the others. He said that relations among the people in the Divisional Managers Meeting had been very steadily improving over the years, and in his opinion their morale was now good enough to allow them to approach their work in a constructively critical fashion. In comparison with management groups in other places of which he himself had been a member, he knew their own morale was extremely high, an appraisal he shared with those others who had also had management experience elsewhere. The quality of their relations, in particular the freedom to speak to each other, was beyond comparison with other situations they had known. This did not mean that things were perfect, but they were able at least to recognize difficulties and to try to deal with them.

Of the reasons for the periodic breakdowns in communications there was some doubt. It was the consultant's impression, derived partly from the interviews, that there was from time to time some diffidence among the divisional managers about making even constructive criticism of each other's work, although they did feel that if they could do so more freely it would be helpful. The main anxiety was apparently about open criticism in meetings in the presence of others, particularly the General Manager, who, it was feared, might take up the criticism of the person concerned. They were all rather sensitive to criticism from him and therefore tended to protect each other, with a consequent diminution of some of their spontaneity in discussion.

If this analysis was correct, it meant that in this higher management group, with its genuinely good relations, it was nevertheless possible to have, co-existent at a somewhat deeper level, an undercurrent of anxiety about criticism—a kind of fear of victimization—of sufficient strength to interfere to some extent with their freedom of communication with each other. That such fears were to some extent at least an accompaniment of the hierarchical structure of the group did not occur to the consultant at the time. We shall return to this topic in the next section.

The Next Rank of Management

In addition to communications within the top management group, information was elicited on top management's view of the efficiency of its communications with the next line of executives. Each divisional manager made an assessment of how much his own first line subordinates were interested in, and in agreement with, the general direction of the decisions taken, and policy laid down, in Divisional Managers Meeting. The General Manager independently made the same assessment.

There was very considerable agreement between the General Manager and the divisional managers in their assessment of the attitudes of the next line management. The Personnel Division, the Service Division, and the second line Works Division were assessed as being neither strongly in nor yet out of agreement with Divisional Managers Meeting policy and activities, and as being more or less content to accept most of higher management decisions without comment. In contrast, the Technical Division was assessed as rapidly moving towards greater agreement and an increasing degree of concern with what went on in the Divisional Managers Meeting. The Finance Division, the Technical Representatives of the Commercial Division, and the first line Works Division groups, which were assessed as least positively in sympathy with some aspects of Divisional Managers

R 239

Meeting policy and activities, were also assessed as most interested in what went on. But, irrespective of the degree of interest or agreement of their subordinates, taken altogether the divisional managers were fairly well in accord in the opinion that their subordinates did not feel either that they made much contribution, or were consulted to any extent, on top management policy.

Because he was unable to get down to some of the more fundamental aspects of these blockages in communications both within the top management group and between that group and the next line of management, the consultant was unable to make much of a contribution towards alleviating these difficulties. But the mere examination of the matter did, by itself, lay a foundation which, as it so happened, was used in subsequent work in strengthening two-way communications with the executive chain.

Works Council Representation and Executive Communications
Many of the difficulties of two-way communication within the executive chain, particularly at top executive level, came out in a rather unexpected manner during the investigation by the Internal Development Committee of the structure of the Works Council. The consensus of opinion in the Divisional Managers Meeting, when the matter was taken up on 20th December, was that the existing arrangement of having all but one of the divisional managers on the Council was a cumbersome one. It would be far better, they thought, to allow for directly elected representation from the three Staff Committees in place of so many senior management members. Top management would be quite sufficiently represented if the General Manager alone was an ex-officio member, with the right to appoint one or two divisional managers to accompany him, and if the divisional managers were allowed to have one elected representative besides.

In these proposals were reflected some of the unresolved issues within the top management group, two of which came out into the open. One was the General Manager's lack of assurance that he would be able to act consistently at Council in a way which would gain full acceptance by his management colleagues. His hope was that if he had one or two of his divisional managers with him, and if they were all in agreement, then they would be acting in a way satisfactory to the rest of management.

240

The other was the implication, in the divisional managers' expressed wish to have their own elected representative, that the General Manager would not adequately express their point of view. Hence the proposals for both appointed and elected divisional management representatives, to act as a kind of system of checks and balances on the General Manager and on each other. They felt at that moment no guarantee that there would always be a sufficiently high level of communication and work-through among themselves to ensure that the General Manager could represent the views of them all.

The matter of top management representation on the Council was the subject of further discussion on the 21st February, when the Divisional Managers Meeting came to consider tentative proposals from the Internal Development Committee for the setting up of a multi-sided Council made up of representatives of all strata in the factory. The Medical Officer (who, it will be recalled, was Chairman of the Works Council) now took a leading part in the discussions. He said he could see good reason why the General Manager should want to have other divisional managers at Council Meetings to advise him, and he thought it would be well to make provision for the General Manager to call on any management personnel, whether divisional managers or not, as occasion warranted. He did not consider, however, that there were any valid reasons why these management members co-opted to Council meetings for advisory purposes should have voting powers. He then turned his attention to the divisional managers, arguing that, if in principle the factory was to have a Works Council elected from the various strata of the organization, with the General Manager in an ex-officio capacity representing the factory as a whole, then surely there was no reason why the divisional managers should have any special representatives on the Council at all, for, after all, as members of Grade I Staff, they would be represented through the Grade I Staff Committee.

These forceful arguments from the Medical Officer, and second thoughts by some of the others, caused the divisional managers to change and finally withdraw their plan to seek special representation for themselves on the Council; but there still remained the question whether or not the General Manager should be able to appoint to the Council one or more of his subordinates to help him overcome the difficulties in communications which they had with each other.   The divisional managers, led by the Medical Officer, argued strongly that the General Manager should be the only non-elected member of the Council.   But he in turn argued just as strongly that he required to have with him at least one other divisional manager with full voting power if he was to be sure of following the wisest course. As one of the divisional managers put it, it was a strange reversal of usual form to have the divisional managers urging the General Manager to represent them, and the General Manager saying he did not feel he was equal to the task.

The consultant, stressing that he could not advise in favour of one course of action or the other, did suggest that one of the factors they might wish to take into account was which course would be likely to contribute most to keeping the channels of communication open, or at least to improving the communications set-up in the management chain.   The General Manager said there was little doubt in his mind that the fact of himself being the sole management representative on the Council would create the greatest demand for improving the executive channels of communication, for good communications would be necessary if other members of management were to have any influence on his views.   With this he began to waver in his attitude, and said he thought he might make an attempt to do the job alone in order to see what would happen, recognizing of course that he could in any case have others with him for advisory purposes.

Their final decision that they would recommend that the General Manager should be the sole voting representative of management, as such, on the Works Council, meant that the divisional managers now had considerably more confidence in their relations with the General Manager and in their capacity to communicate with him, and so influence his views as to arrive at mutually satisfactory decisions.   Equally the General Manager's acceptance of this task, at least for a try-out, represented an increasing feeling of confidence in his capacity to maintain effective two-way relations with his divisional managers.   Although there were possibly other factors implicit in the decision as well, such as satisfaction from pressing the General Manager towards a position of which he was somewhat afraid, these did not negate the positive aspects.   The consideration of communications problems within the top management group, in terms of their own working relations as executives, had had some effect, even though the difficulties were by no means completely worked through.

CONCLUSION

This chapter has described the attempts of the top management group to resolve difficulties which had their roots in the splits in the executive system which occurred a number of years before, when there had been dramatic shifts in the top leadership of the firm. Having decided to sort out these unresolved problems, and to pursue the matter of human relations within their own group as well as in the rest of the factory, they found themselves having to deal first with the relationships between the General Manager and the divisional managers. To achieve this end, they made use of transitional situations such as the Works Council, and transitional figures such as the consultant, by means of which to explore and test out their attitudes, before considering directly their own immediate situation. Having succeeded in tackling the problem of the leader-subordinate relationship in the group, they found themselves up against, and recoiled from, what seemed the far worse problem of relations among the divisional managers. Unable to face this problem directly for the time being, the group terminated the special evening meetings, but continued to have the consultant attend their regular business meetings. In this setting, the working-through process was carried on at a more controlled and steady pace.

The acceptance by the General Manager of his responsibility for the executive leadership of the top management group was accompanied by the giving up of notions that to give orders was necessarily autocratic and hence bad, both ethically and in the sense that orders automatically produced a reaction of rejection. As the General Manager assumed more fully his leadership mantle, his subordinates in turn accepted theirs. With their executive duties more sharply in focus, they were enabled to take the lead in building a Works Council more broadly representative of the factory, and hence giving greater sanction to their own executive authority. On this Council, the divisional managers felt sufficiently secure to have their managerial interests catered for solely by the General Manager. The working-through of relations in the group had gone sufficiently far for inflexible desires for perfect leadership to be relinquished, and for fallibility to be accepted and mistakes allowed as a necessary and useful part of the behaviour of the General Manager in exploring new paths of development.

The New Series of Special Evening Meetings

In October, 1950, the Divisional Managers Meetings initiated a new series of special evening meetings, and in so doing showed themselves to have come round in a spiral to a point similar to, but at a level above, the point where they had left off the first series of meetings in 1948. These meetings take place every week following the regular business meeting, last for two hours, and are official executive meetings with the General Manager in the chair. The first topic for discussion has been the policy covering the relative authority of the functional and the line managers; a topic which is a part of that question of the relationships between the divisional managers which led to the temporary cessation of the original evening meetings. The group's capacity to tackle such problems has undergone a considerable change; there being a marked diminution in the amount of anxiety about accepting the fact that, although this particular problem is one of widespread concern in industry, they will find a solution satisfactory for themselves mainly by considering their own relationships, without seeking refuge in pleasing and neat administrative generalities.

A detailed account of these discussions cannot here be given, since they are currently proceeding; but one point which can be mentioned (that to do with functional relationships) will serve to indicate the present capacity of the group to deal directly with what may ordinarily be considered very delicate matters. It will also throw further light on one of the factors which made the problem of rivalry and competition such a thorny one.

The explicit and emphatic policy of the firm was that every man should have only one executive superior and should have delegated to him suitable authority for carrying out his duties. Examination of actual practice revealed that, in two important respects, executives were subject to checks from managers besides their own superiors; if a manager who had the job of providing services for others not under his command felt that a given request for service was contrary to policy, it was his responsibility to withhold service pending appeal to higher authority; and, if a manager in an advisory role had his advice turned down, it was his responsibility to report this to his own superior. Discussion of these discrepancies between the policy as professed and as practised soon led to the realization that one main source of the problem lay right within the top management group itself. Such an insight was now acceptable. The General Manager could allow himself to recognize his anxiety about giving unreserved authority to each of his divisional managers to carry out jobs he delegated to them. He was able to discover that he had unconsciously (despite his most earnest attempts consciously to implement his views about clear-cut delegation) played safe, by having his divisional managers informally check upon each other. This informal checking meant that the divisional managers had to repress their own differences more intensely. For not only were there the ordinary rivalries and other attitudes to be found within any group of subordinates, but they were encouraged to concern themselves with the business of their colleagues in a manner which appeared very secret, in view of the professed policy by which each was responsible only to, and watched over by, the General Manager.

When these features were brought into the open, there was hesitation, worry, and distress which lasted over a number of meetings. But gradually fears were brought under control. The fact that members of the group had for some years felt guilty about reporting on each other (even though only informally in conversation) was talked about for the first time. Considerable relief ensued. It became possible for each person unequivocally to state his personal views on the policy issue under consideration, despite the fact that their deliberations would decide how much control, if any, they would have over each other's work. The differences in outlook which emerged could be related to differences in past experience, in personality, and in the roles occupied. Some held that full responsibility should be delegated to one person, others that it was essential to assign some kind of watching and checking brief to managers other than the immediate executive superior of the manager concerned, and a few maintained positions somewhere in between. The solution finally adopted was that managers should be responsible only for their own subordinates, and should have no checking or watching responsibilities for others not under their own command (see Appendix Two). In arriving at this decision they were able to take the personal differences and preferences of individual members into account, since their tolerance for such differences had grown. Because they have been able to take these personal factors into account, along with the demands of their organizational situation and their knowledge of administrative practice, it is likely that they have arrived at a solution which individually they will be able to apply with consistency, and which they will be able to modify and change as necessary.

The constructive social developments which the top management are trying to achieve in the factory as a whole, depend not only on the quality of their own group relationships, but also on the quality of their leadership of their own immediate subordinates. Their aspiration in this latter respect is illustrated by a recent change in terminology: the Divisional Managers Meeting is now called the General Managers Meeting, and this is being followed by similar changes throughout the executive system. The practice of each manager regularly meeting his own direct command as a group is spreading; a development which was described in the case of the change from the Superintendents Committee to the Works Managers Meeting. Such changes in terminological emphasis would not necessarily be significant in and of themselves; the important point is that in this case they are an accurate reflection of changes in outlook and behaviour which have already taken place.

PART THREE

ANALYSIS OF CHANGE

The main themes arising from the foregoing background data and case histories will be drawn together and analysed; the goal of the analysis being to throw light on the processes of joint consultation and executive action and the relation between them, as well as on the dynamics of the changes which have occurred in the factory, and their meaning in terms of social adaptation. Finally, a summary is given in the form of a general interpretation of the broad movements in Glacier's history brought about through the interplay of technological and social forces.

CHAPTER NINE

DEFINITIONS

Some Terms Necessary for the Analysis

THE method of analysis to be used in the following chapters will be to study how the pattern of social activity at Glacier, as described, has come about through the interaction of the firm's organizational structure, its customary way of doing things, and the behaviour of its members. That is to say, we shall study the interaction of social structure, culture, and personality. But before undertaking this analysis we shall define the terms to be used throughout. In those instances where these definitions are found to coincide only partly with current usage, this is due to the fact that no consistent and widely accepted popular or scientific usages could be found. Other concepts applying to particular aspects of the analysis will be introduced and defined as necessary in the appropriate sections.

SOCIAL STRUCTURE, ROLE, AND ROLE RELATIONSHIPS

The *social structure* of Glacier is the more or less recognizable and stable organizational pattern of the firm including its executive hierarchy, consultative system, and grading system. Structure in this sense is made up of a network of positions which can be occupied by individuals or groups; these positions being defined as *roles*. Thus, for example, the social structure of the executive system is made up of the network of roles of general manager, divisional manager, down the chain to the role of operative, patterned in a hierarchy of superiors and subordinates. Structure and roles are felt by the individual to be external to himself, as something "out there", and it is by the process of taking up a role or roles in the social structure that the individual becomes a working member of the community. The nature of the social structure lays down the formal relations between roles, or *role relationships*; for example, the role of the superintendent in a given department is linked to the role of works manager, and to the role of foremen and section supervisors in the department, in a specified way. The implication of formal role relationships is that a person taking a particular role must establish relationships with certain other specified individuals occupying related roles, not as a matter of choice but through the demands of the social structure.

The pattern of roles constituting social structure links living persons with organization, and also links the people in an organization with each other. But roles are not people, and people are not roles; a point which, for reasons to be examined later, caused some confusion at Glacier. People carry a multiplicity of roles, a fact which is well recognized in ordinary life, where an individual may carry such diverse roles as father, club member, sportsman. One person may also carry a number of different roles at work, that is to say, he may occupy more than one position in the organization; conversely, one role or position may be split up and occupied by more than one person, for example in co-delegation. The significance of having a clear perception of the organizational structure of a concern lies in the fact that structure defines roles and role relationships. To roles are attached job responsibility and authority, and role relationships determine the official relationships possible between people, the pattern of roles and of relationships giving the formal setting in which working behaviour takes place.

The social structure of the firm can be seen from one point of view as being made up of horizontal levels or strata, as for example in the organized constellations of roles such as Grade II Staff, and the more diffuse groupings such as the management stratum or the skilled worker stratum. Values are placed both upon roles and people, some being more, or less, highly regarded than others: the value attaching to a role or stratum of roles will be referred to as its *status*, and that attaching to a person as his *prestige*. The status attaching to a role or a group of roles may change without any change in organization, and more than one distinct status may be assigned to the same role or stratum by different groups in the concern. The regard held for a person in a given situation is a combination of his prestige and the status of the role he occupies, while conversely the status of a role may undergo change due to the prestige of the person who occupies it.

CULTURE

The culture of the factory is its customary and traditional way of thinking and of doing things, which is shared to a greater or lesser degree by all its members, and which new members must learn, and at least partially accept, in order to be accepted into service in the firm. Culture in this sense covers a wide range of behaviour: the methods of production; job skills and technical knowledge; attitudes towards discipline and punishment; the customs and habits of managerial behaviour; the objectives of the concern; its way of doing business; the methods of payment; the values placed on different types of work; beliefs in democratic living and joint consultation; and the less conscious conventions and taboos. Culture is part of second nature to those who have been with the firm for some time. Ignorance of culture marks out the newcomers, while maladjusted members are recognized as those who reject or are otherwise unable to use the culture of the firm. In short, the making of relationships requires the taking up of roles within a social structure; the quality of these relationships is governed by the extent to which the individuals concerned have each absorbed the culture of the organization so as to be able to operate within the same general code. The culture of the factory consists of the means or techniques which lie at the disposal of the individual for handling his relationships, and on which he depends for making his way among, and with, other members and groups.

PERSONALITY AND PERSONALITY DISTRIBUTION

By *personality* we distinguish the total psychological make-up of the individual, his attitudes and beliefs, desires and ambitions, likes and dislikes, capacity for making relationships, leadership and technical skills and ability, intelligence and wisdom. While many aspects of personality are conscious, in the sense of being known to the individual and apparent to others, many of the most important are not. People are unaware of many of the basic reasons why they behave in the way they do, and to discover these reasons is difficult and time-consuming, and calls for special procedures. The importance of these unconscious factors in human behaviour is that people, unknown to themselves, can subscribe to numbers of opposing and inconsistent beliefs, and be driven by conflicting motives, some of which are conscious, and some not. It is these multiple beliefs and motives which give rise both to the commonly observed inconsistencies of human behaviour and to the fact that stated policies are not always easy to carry out in practice, no matter how strongly they are consciously endorsed and believed in.

Manning the social structure for the factory as a whole, by filling the available jobs, requires a wide variety of types of people with a wide variety of skills and interests. The relative frequency of distribution of types of people in the concern will be referred to as the *personality distribution* of the organization. The screening of newcomers, training of personnel, selection procedures for promotion, are techniques for getting a personality distribution which matches the practical requirements of manning the social structure.

THE INTERACTION OF STRUCTURE, CULTURE AND PERSONALITY

It is its particular pattern of structure, culture, and personality that gives the factory its own unique character, and this pattern is continuously being modified and developed, due to the interaction of changes in structure, culture, and personality. The production requirements are constantly undergoing change, owing to external economic and social forces; the personality distribution is constantly changing as the result both of changes in individuals and of the coming and going of personnel; and culture undergoes continuous modification as a result of new techniques and ideas brought in from outside. Frequent small changes occur in the course of day-to-day adjustments in procedures, and occasional large-scale changes are made necessary by accumulations of problems.

The personalities of individuals affect social structure in the sense that a person is never perfectly suited to a role; gross discrepancies between personality and role being handled not only by redeployment or personality readjustment, but also by the individual gradually adapting the role to suit his own needs or by those in charge making an explicit alteration of the role to suit the available people—a procedure which usually entails modification of other roles as well. On a larger scale, when the personality distribution of the members of the organization is such that there are no available people to fill a vacancy, either new people have to be brought in from outside, or else the structure of the organization must be altered to suit the available manpower. Changes in people in turn alter the culture of the organization; the taking on of new personnel carrying with it the introduction of cultural techniques which the new members bring from their previous organizations.

The effects of changes in culture on personality and on structure can be shown in the case of the introduction of new production methods, as, for example, mass production techniques. Such techniques usually mean breaking down the complexity of jobs and, hence, require a personal readjustment on the part of the individuals concerned, and a change in the type of personnel taken in. At the same time new and different kinds of planning staff are required, and thus the social structure must be reorganized to make a place for new departments and new roles.

Changes in structure may occur, to take one example, as a result of an increase in the size of the organization so that additional layers have to be introduced into the executive hierarchy and the shop floor is placed farther away from top management. Such changes make considerable demands on the individuals and frequently call for the display of new skills and abilities and the perception of new role relationships, as in the case of the new types of relationship required by the introduction of specialist departments. Equally, cultural readjustments are required, in the sense that different procedures and codes are necessary for maintaining satisfactory relations as the extent and complexity of the executive structure either increases or decreases.

The life of the factory is thus an incessant interaction of structure, culture, and personality; changes in any one area requiring changes in the others. Stresses arise when changes occurring in one area are met by resistance to change in another; for example, a change in structure may require that certain people take up new roles to which they find difficulty in adjusting themselves. In planning changes of any type, therefore, it is necessary to pay attention to the requisite alterations in social structure, the likely shifts in culture, and the personal readjustments of the individuals concerned.

POLICY: OBJECTIVES, PRINCIPLES, AND PROGRAMMES

The *objectives* are the goals or long-range purposes of the concern, which have been set out in the Company Policy document (Appendix I). The *principles* are the elementary or fundamental ideas governing the running of the organization. These are ordinarily not explicitly formulated, but as examples could be quoted the belief that people will behave responsibly if treated as responsible, or the principle of the ultimate social value of efficient production methods and of mass production. The *programme* of the organization is the outline of strategy and general procedures for implementing principles and reaching the objectives of the concern. The *policy* is the totality of objectives, principles, and programmes, governing the running of the firm in all its aspects—technical, financial, commercial, administrative, and social. Its policy is part of the culture of the concern, and is a function of the way the people concerned behave, as well as the things they say, whether verbally or in writing.

RESPONSIBILITY, AUTHORITY AND POWER

As used in this study, responsibility and authority are qualities of the social structure of the organization, and power is a quality of the individuals or groups who occupy positions in the organization. The *responsibility* attached to a role is the sum total of tasks, people, and equipment given into the charge of a person by virtue of his occupying a given role. For these he is answerable, in the sense of having to give an account to his superior, who in turn is responsible for him. Subsumed under responsibility are the duties and obligations which an individual undertakes as part of the process of being given authority; responsibilities which are not satisfactorily balanced by the authority necessary to discharge them being felt as a burden. The *authority* attached to a given position in the firm is the statement of what any person (or body) occupying that position can do, whom he can instruct, what equipment he can use, and what he can authorize to be done. In short, the authority system of a community is a formal structure which defines and regulates the means and directions in which individuals or bodies may exert power. In contrast to authority as an attribute of a position, *power* is an attribute of an individual or group; the term defines

254

the strength or intensity of influence that a given body or individual is potentially capable of exerting at any given time, regardless either of the role assumed or the authority carried. Power is a product of such factors as skill and knowledge, capacity to lead and influence people, and physical and intellectual strength and group cohesion. Power and authority can vary independently of each other; thus, for example, a subordinate, through greater technical skill, may in some respects have more power than his superior, although by position he has less authority; or an unauthorized strike committee may carry greater power than the formally authorized trade union leaders. It is an essential of sound organization that power and authority must not be too disparate. This suggests that individuals and groups should occupy positions and have responsibility in the organization commensurate with their power. When power is too much in excess of responsibility and authority, an explosive situation develops; when power is inadequate, authority will be discredited.

THE SANCTIONING PROCESS

As we shall use it here, the full sense of the word "sanction" emerges if we consider its two opposite but interconnected meanings, depending on the verb which accompanies it; to *give sanction to*, and to *apply sanctions against*. In our definition, to give sanction means to permit authority to be attached to certain roles and to accept the use of power by persons or groups occupying those roles. In order to give sanction in this sense, the sanctioning person or body must have the power to impose punishment—whether by boycott, obstruction, penalizing, strikes, or complete destruction—that is, to apply sanctions against the person or group in the role whose authority has been sanctioned, or to support that person or group in imposing punishment against any other person or group where there have been infringements of accepted and sanctioned activities and behaviour.

In this sense, the sanctioning process is a cultural mechanism by the use of which power is linked to authority. It covers the relatively stable and socially accepted process, such as the more or less fixed sanctioning of roles like the Speaker in Parliament; the more evanescent processes, such as the temporary giving of sanction to a specially elected delegation to act on behalf of a group on one specific issue; and the less easily recognizable but none the less real processes such as public opinion. The *degree of sanction* given in any particular case will differ according to how many of the various bodies concerned give sanction, and how much of their total power each of these bodies is willing to invest in support of the sanction they give. Complete sanction can be conceived as being whole-hearted support from all concerned, and partial sanction being anything less than this. *The strength of sanction* will vary according to the power of the sanctioning groups, so that partial sanction from a powerful group may be more useful than complete sanction from a powerless group.

CHAPTER TEN

THE SANCTIONING OF AUTHORITY

Joint Consultation and Policy Making

THE sanctioning of executive authority may be thought of in respect of structure and of people. As regards structure, sanctioning is required both for the authority attaching to specific roles, and for that belonging to the executive system as a whole. We shall refer to *role sanctions* when single positions are involved, and to *system sanctions* when reference is rather to a total system. Role and system sanctions are of two kinds, those given from outside the organization (e.g. by shareholders and customers), and those given from inside (e.g. by the nature and demands of the work task, and by the consensus of opinion achieved through joint consultation). With respect to people, there is sanctioning of the authority of particular persons in their executive roles which will be referred to as personal sanctioning. *Personal sanctions* are given to the individual by the approval of superiors, the co-operation of colleagues, and support of subordinates, and also by himself through his own personal capacity to tolerate and accept responsibility and authority.

In this chapter we shall consider the role and the system sanctions, or how executive authority is sanctioned in the concern as a whole, leaving to the following chapter the question of personal sanctions, or how the individual may get sanction as a fit person to exercise the authority necessary to carry out his scheduled tasks. By taking into account the main sources of sanction both inside and outside the organization it will be possible to examine joint consultation as one facet of a more general sanctioning process, and to show how the power of the various sanctioning groups is regulated in the Works Council and balanced by the unanimity rule.

SYSTEM AND ROLE SANCTIONS GIVEN BY SOCIETY

The factory operates within its larger society. It is successful as an industrial undertaking in so far as it succeeds in maintaining a connection between the new methods it is trying to develop and the central features and trends of the culture of this society. The more the members of the concern intuitively understand, and can also consciously put to use, the established practices of British culture before seeking new practices, the greater their freedom to experiment. The more deeply they are steeped in accepted ways, so much greater will be their skill to move along new paths without setting up stress between themselves and their society. A self-sanctioning industrial organization is impossible; conformity with the larger society being enforced in innumerable ways, some of which we shall deal with here. Intelligent and knowledgeable conformity brings external sanction and a degree of freedom. These external sanctions are to some extent carried inside the organization in each individual member of the concern, but the focus of our attention will be on the extent to which they are carried in a recognized and defined way by the Board of Directors, the top management, the specialists and technicians, and the trade union representatives.

Economic Sanctions and the Company's Product

The primary sanction for the work of the organization derives from the nature of its task and the product it makes. This sanction is given by investors who provide the capital for the company, and by the customers who buy its products. Without sanction from both these sources, expressed by means of financial payment, the company in its present form simply ceases to exist, and hence the groups inside the organization to whom sanction is given from these sources carry exceedingly great authority. How far the authority of the executive system is sanctioned by investors and customers depends on the profitability and potential profitability of the concern, and on the quality of its product.

The Board of Directors carries responsibility for the money invested by shareholders; each member of the Board being held personally liable in the eyes of the law for the manner in which he discharges this responsibility—a liability which cannot be delegated to other members of the organization. It is consistent with this special responsibility, as laid down in the Memorandum and

Articles of Association of the Company and in the Companies Act, that the Board should be endowed with full authority by the shareholders to run the business in conformity with company law, in between annual general meetings. It is the Board which is authorized to spend money, and to have regular access to the profit-and-loss position. Inside the organization the authority of the Board is carried by the Managing Director, to whom is delegated the responsibility of managing the concern between Board meetings.

The sanctioning power of the customer is transmitted directly (rather than through the Board) to the General Manager and to the Commercial Division, who have direct contact with customers. Customer sanction has put a halo of special authority round the commercial section, whose members can commit the firm to deliver specified goods in a specified time at a specified price. But this authority was not until recently systematically organized in the executive system. It was relayed into the factory through the Production Engineering and Production Control departments. By virtue of their customer-derived authority, these departments were able under old policy to exert so large a measure of control over rates, production schedules, and methods, that those in the shops, supervisors and workers alike, came to feel that they had more than one boss. Under the new policy, authority sanctioned by the customer is carried by the General Manager himself, who then delegates to the various divisions their more specific responsibilities in relation to it: to the Commercial Division those of making sales contacts and of entering into contracts with buyers; to Production Engineering and Production Control those of building layouts and of setting prices and schedules; to the Works Division the job of getting out the work efficiently and of deciding whether or not they are able to follow the methods, prices, and schedules drawn up by other departments. By means of this change in policy, customer sanction is now carried formally and explicitly down the executive chain from the General Manager rather than in the concealed, crosswise manner which was the case before.

Political and Governmental Sanction

The factory must adhere to the laws of Britain and implement governmental regulations, trade agreements, negotiated wages agreements, and conditions of work. The top management members carry the responsibility for following legal and governmental regulations regarding trading and manufacturing methods and with it the authority for implementing the actions required. The trade union representatives or shop stewards inside the factory are responsible to their unions for seeing that the conditions of employment conform to the minimum standards established by the Confederation of Engineering and Shipbuilding Unions in negotiation with the Engineering and Allied Employers Federation. This responsibility gives the shop stewards authority to meet together and to make direct contact with top management on matters affecting union members. This internal system is supported by the direct and authorized relationship between shop stewards and district union officials and district union officials and management. The extent of the responsibility and authority of the shop steward depends upon the extent of trade union organization of the members of the concern, the present more comprehensive organization carrying with it not only increased authority but heavier responsibilities as well.

Technical and Professional Sanctions

The position with regard to the authority of many of the specialists and technicians is complex and not clearly defined. They carry responsibilities to various outside professional bodies with regard to the conduct of their work, and tend to be invested with a tacitly recognized authority by virtue of their membership of these bodies, regardless of their own particular skills and ability. For example, the medical officer acquires special responsibilities and authority by virtue of his membership of the medical profession. On many matters of prevention of industrial disease his word is unquestioned. This combination of responsibility and authority is likewise characteristic of other specialists and technicians, such as production engineers, metallurgists, and engineering research specialists; but the operational involvement of their roles makes the boundaries of their specialist authority less clear *vis-à-vis* general management.

Sanctions from Community Standards

Members of the concern, having been brought up within British culture, are unconsciously limited in the extent to which they can deviate from its traditional practices. Moreover, patterns of behaviour and organization within the concern must remain consistent with the patterns of behaviour and expectations of a community that carries this culture. Too wide departures would rapidly bring into operation those subtle forces which cause an organization to become unacceptable and to be rated low on the scale of places where employment is sought. The management may exercise the authority granted to it in the prevailing industrial culture only so far as it accepts the obligation of conforming to the standards expected. In that part of the London region in which the firm is situated, more so than in some areas, these standards include the idea of treating employees as human beings rather than as cogs in a machine, and of providing adequate welfare and personnel facilities, while imposing a discipline that is not too rigid.

The Contract

The mutual acceptance of responsibility and authority by the firm and its members is expressed in the contract of employment —covering the conditions of work and remuneration, the rules and regulations to be observed, and the work to be done. In signing his contract the individual sanctions the authority of the firm to ensure that he fulfils his obligations. The contract at Glacier does not exist as a single document. It is a combination of the conditions of pay and work established in the initial interviews and set out in the Engagement Form, the rules and regulations laid down in the Rule Book, policy as stated in the Company Policy Document, and unstated procedures, some of which belong to the generally accepted industrial code, others to the culture of Glacier itself. In this sense the full meaning of his contract does not become evident to a new employee until he has had the opportunity of being with the firm for some time and of assessing its worth in terms of his own personal needs. During this time the firm also has the opportunity of assessing how far he, as an individual, is likely to fit in.

If the members of the factory generally are dissatisfied with the contract, they may try to change it. If only isolated individuals are dissatisfied, they may leave. Conversely, authority is vested in managers to discharge from their own commands individuals who do not carry out their responsibilities. With the General Manager rests the final authority to discharge individuals from the organization. The appeal mechanism provides safeguards against abuses of this authority.

THE SANCTIONS FROM INSIDE THE ORGANIZATION

Two sources of authority inside the organization will be considered: those sanctions arising from the technological requirements of the task of producing and selling bearings in a competitive market; and those given by consensus of opinion expressed through the consultative mechanisms.

The Task-Sanction

In dangerous occupations, where there is an imperative need for discipline, the sanctions imposed by the work task are self-evident. The work at Glacier involves equally real, albeit less immediately apparent, needs for technologically expressed authority. The types of material used, the degree of precision required, the extreme specialization, the jobbing and multifarious nature of the assignments, are characteristics, inherent in the production of bearings, which set limits to what can be done, and give special authority to technological roles. Instructions accepted as necessitated by technological requirements carry greater authority than those which are not. The sanctioning power of the work task lies in the fact that if the demands imposed by the nature of the task and the materials used are not recognized and accepted, then production suffers.

Sanction deriving from the nature of his task is carried by each person in his working role. But recognition of this task-sanction is expressed in the organization as a whole through the executive system. Ultimately, top management is in the best position to distinguish what is technically possible, having regard to problems of sales, costs, production, organization, and morale. Task-sanction passes up the executive chains from all parts of the factory to the general manager who co-ordinates the factory's collective work experience, and delegates necessary instructions and authority back down the executive chains. The hierarchical organization of authority has its basic sanction in the task; the differentiation of work, in such a way that one person can delegate parts of his responsibility and authority to others under his control, being sanctioned mainly by the greater efficiency made possible. Conversely, the sanction automatically disappears when hierarchical organization no longer achieves greater efficiency than would be achieved by the members of an executive system working independently.

Democratic Sanctioning through Joint Consultation

Democratic processes operate directly in some of the general sanctions given by society at large, in the sense, for example, that laws are democratically arrived at. Inside the factory, democratic processes that conform to political tradition find their expression in joint consultative mechanisms. These mechanisms have widened the scope of the sanctions given to those who occupy positions of authority, by providing them with the sanction of subordinates as well as of superiors. The degree of sanction given by consultative mechanisms has been substantially increased by the modification in constitution which gave elected bodies authority to take part in the formation of policy. The complement of policy-forming is checking on the execution of policy, and in this particular the consultative system acts as a channel through which may flow any general dissatisfaction with the manner in which authority is used. On the one hand, behaviour may be assessed as in keeping with, or contrary to, agreed policy, and on the other, policy itself may be changed or clarified, or more effective implementation may be demanded.

The nature of the interaction between authority and democratic sanctioning displays itself in the paradoxical relation between the scope of function of the consultative bodies and the strength of the executive. With democratic mechanisms through which all members are enfranchised, in the sense of having a say in policy, the authority of those in managerial roles is both questioned and yet upheld by those whom they control. The more far-reaching the control exercised by the consultative system, the more complete the authority invested in the executive system. The new consultative system at Glacier may be regarded as an attempt to discover how far the extension of democratic rights inside an industrial organization can assist the ability of management to manage.

THE REGULATION OF POWER

The change in structure of the Works Council completed the differentiation of the executive from the consultative chain; the General Manager being left as the top level channel of communication between the two systems, with full responsibility for implementing the views of the Council in management, and for placing the views of management before the Council. The Works Council, as an elected body representative of the total factory, acquired policy-making functions as a member of a company-wide policy-making network composed of three groups. These are the Board of Directors, for the company as a whole, and the Works Councils of each of the two main factories. Company policy can now be decided only by unanimous agreement of all three. The way in which this unanimity procedure works in practice is very much a matter of the balance of power between these three groups, and the special interests they represent. Unlike responsibility and authority, which are structurally defined and relatively constant, the relative power of these groups shifts continuously. Before the significance of the unanimity rule can be understood, the sources of power of each of the three bodies must be examined.

Power of the Branches of the Policy-making Network

The present Board of Directors owes much of its power to the fact that a large interest is concentrated in the hands of a few shareholders who consistently support its policy. Other sources are its special knowledge of the general financial situation, and its own group cohesion. The particular composition of the Glacier Board—five whole-time executives out of a total of seven directors—also affects its power, increasing that attendant upon intimate knowledge of the firm, but lessening that arising from wide outside experience. The presence of the Managing Director and four divisional managers makes it difficult for management to scapegoat the Board; but there is again a loss of power in this same situation, due to the difficulties of keeping roles distinct. For the Board to influence the Managing Director requires the Managing Director and the four divisional managers to forget their day-to-day working roles, and to adopt an outlook in which, as members of the Board, they are superior to the Managing Director and responsible for instructing him on policy. For the Managing Director, this means metaphorically stepping outside himself, and scrutinizing his own work in one role from the viewpoint of a position higher up; and for the divisional managers it means adopting an attitude of seniority towards the Managing Director in the presence of the very person who occupies that role but who is, in his turn, once again senior to them as Chairman of the Board. We shall consider in more detail in Chapter Twelve some of these problems related to multiple role carrying.

The power of management depends on technical expertise, skill in leadership, the efficiency of operation of the executive communication chain, and the cohesiveness of the management group. The top management of each of the two factories, through the general managers, and finally the Managing Director, have a say in policy-making with widespread effect, and the shop floor and junior management can only speak executively through them. This does not necessarily mean that the executives nearer the top have greater power, this being a matter of how well organized are the lower levels, but it does mean that the higher level managers have access to greater stores of information, and these do confer power.

The power of the elected bodies in carrying out their authority sanctioning task, and in particular the power of the Works Council, which has the authority to participate directly in policy making, derives from the skill of the elected representatives themselves, as well as the degree of organization and cohesiveness of the members of their constituent groups. As a special case, the new Works Committee also acquires powers from its relationship to the trade union movement; the shop stewards being able to get support and advice from union and district officers, as well as direct support from union officers where negotiation is involved.

Power and the Principle of Unanimity

The question, "Where in such a set-up does the final authority for deciding policy rest when unanimity cannot be reached?" is one that has no absolute answer. Where legal issues are involved and the directors are held responsible at law, the Board of Directors will usually have to make final decisions. Apart from cases of this type, real failure to reach unanimity means that the matter will be decided by power. But whereas the authority and the responsibility of the different bodies are known and defined, their power is constantly undergoing change, and therefore the main weight for final decisions on policy is likely to rest with different bodies at different times. For example, the relative power of the Board of Directors and the trade union leaders under conditions of unemployment and weak union organization, will be sharply different from that obtaining under conditions of full employment and strong, coherent trade union organization. It is precisely because power relations may shift, while the authority structure remains unchanged, that the firm's unanimity principle is of such value, for it allows the continuous testing-out and exploration of the power situation by means of constructive discussion, instead of the intermittent testing of power which accompanies executive policies and actions which have not been agreed and which when unacceptable lead to a piling up of stress and to explosive outcomes.   Not that the unanimity principle automatically solves questions of power relationships; rather is it to be seen as a mechanism for facilitating more constructive relationships and ensuring more realistic compromises when the necessary motivations and skills exist in those concerned.

The unequal distribution of responsibility, authority, and power in the three groups is associated with differences in the value and extent of their respective contributions to policy in particular spheres.   On the Board of Directors and the top managements falls the main weight of responsibility for initiating financial and commercial policy and technical development. But proposals in these fields are subject to modification in the light of their effects on factory personnel, as perceived by elected representatives.   For example, a programme of technological development which would radically alter methods of production could only be planned by the Board and by higher management in conjunction with the firm's technicians.   But it could only be adopted as policy after consultation with, and sanctioning of the plan by, the Works Councils in the light of an assessment of possible effects on the firm's personnel.   Conversely, on matters such as promotion, methods of payment, status, and other aspects of group relationships, the Works Councils are centrally placed to deal with whatever problems may be thrown up either by the management or by representatives, and to adumbrate principles to deal with them.   Any such principles would, however, affect financial policy, so that the other members of the policy-making network require to have their say.

The delegation of policy-making to a network of bodies, with primary responsibility in various areas falling on different bodies and with final implementation being the subject of unanimous agreement, is a pattern which comes to grips with many real problems.   It begins by recognizing that members of the Board of Directors or of the Works Council may be better equipped by training, position, and experience to deal with some problems than with others. It takes into account the fact that changes in any one region of policy have effects on every other region and, hence, that effective joint consultation means consultation on all matters between all the bodies, though not necessarily in the same order or for the same reasons. The operation of the procedure may be exemplified in the general policy document appended to this book. Initiated and drawn up by the Managing Director with the help of the higher management of both factories, it was issued by the Board of Directors and approved by the Works Councils as a working draft of company policy. Final approval has yet to come, and will be accorded only after the document has been considered in detail in both Works Councils and at all levels of management. In the course of this scrutiny, modifications are being made, and these must be ratified by all concerned.

The members of the factory recognize that this unanimity rule may lead them into situations of stalemate. But they prefer to maintain it on the grounds that decisions so arrived at have the best chance of being both the most correct and the most acceptable. Their experience is that so long as group relations remain satisfactory, no stalemate occurs. People show themselves to be flexible enough to modify their views. On occasions when stresses between groups appear, the unanimity rule is still useful. Even should a stalemate occur, it is by this means that the unfortunate consequences are avoided of taking decisions without full agreement, for decisions of this kind are usually impossible to carry out successfully.

Significance of the New Works Council Structure

The original two-sided (management-worker) Council reflected the general split in industry between management and workers. But as the Works Council and the consultative machinery generally began to assume a role in the policy-making structure, a two-sided mechanism became confining. The worker-management Council represented directly by election only some 800 of the approximate 1,400 members of the factory. The other 600 were represented by the members of higher management appointed by the General Manager.

In principle at least, the new multi-sided Council lays the basis for an integrative rather than a split approach. The General

Manager, representing the total managerial structure and indeed the total interest of the factory, meets with elected representatives of all strata. A structure of this kind provides the opportunity for major issues of policy affecting the whole factory to be considered, and for the decisions taken to be implemented by the General Manager through the executive chain. All the forces sanctioning authority in the factory meet in the Council. The General Manager carries the sanctions from outside given by the Board of Directors. He also carries the executive or task sanctions given by the Managing Director and by the managers and other members of the factory in their working roles. The elected representatives carry the sanctions given by their constituents, and the Works Committee representatives those also of their trade unions. All members carry their own personal interpretation of the sanctions given in the general code of industry and in the unwritten culture of the firm. These forces are all at work in the Council meetings. The work of the meetings continues until a point of equilibrium is found which balances incompatibilities between the needs of the various sections of factory personnel, the demands of the task, and the standards of outside society. The new Works Council structure allows more of the real social forces affecting the factory to impinge directly upon each other than did the previous structure with its more limited representation.

The shift to a multi-sided Council required a great deal of give and take among such groups as the different grades of staff and the trade union officials, and a willingness on their part to co-operate in new experiments. As regards trade unions, those sections of the factory which are sufficiently organized are represented by shop stewards, but those sections that remain unorganized are not to be excluded from the consultative machinery. The main problem exists at present in the borderline group—the Grade III Staff—which is partly organized and partly unorganized. A temporary solution has been found by including the one completely organized section of Grade III Staff, the draughtsmen, in the Works Committee for all purposes of representation. This solution has not satisfied the supervisory members of Grade III Staff, who believe they belong with management since they have the command of men below them, yet remain closely identified with the workers in outlook and group membership. Many supervisors were themselves operatives not long ago. The problem of settling where the supervisory grades "belong" illustrates the difficulties encountered when joint consultation is considered solely as a relationship between management and workers. It was possible to sidestep this problem so long as the Works Council comprised only top management and workers. But once the intermediate grades of personnel were brought in, the full complexities of the situation had to be faced squarely. There was rapid recognition that joint consultation is a process whereby the senior executive in the organization as a whole, or in some part of it, meets the elected representatives of all personnel under his command, rather than merely the elected representatives of the hourly-paid workers.

The new Works Council thus provides a setting for straightening out relations between the General Manager, representing all factory personnel in their executive roles, and the elected members, representing these persons touching their rights as individuals or their interests as sectional groups. It also gives opportunities for resolving difficulties between different levels of personnel through the direct contact which their representatives experience with each other. Should a given problem not affect the whole factory, provision has now been made in company policy for sectional consultative meetings between the General Manager and any of the Staff Committees or the Works Committee. Decisions taken at these meetings affect only those concerned, and are subject to review by the Works Council. The new structure is capable, therefore, of taking into account splits such as that between the management and the workers, or between the management and a particular grade of staff, when and as these splits occur, without harm to the multi-sided character of the Works Council itself in dealing with factory-wide affairs.

ELUCIDATION OF POLICY

Those aspects of the culture of the firm having to do with the operation and sanctioning of authority and with the organization and functioning of the internal democratic mechanisms may be summed up under the headings of the firm's policy, including objectives, principles and programmes. Policy—except for the principles, not so far teased out and made explicit, on which the firm's general outlook is based—is being set down in writing.

Drafting and Agreeing the Company Policy Document

The task of setting out a first draft of an explicit statement of company policy, which was undertaken by the Divisional Managers Meeting, was found to be exceedingly difficult. Many practices had grown up in the firm during its early years which were hard to recognize as policy since they had become so much a part of second nature; and even where documentary records existed, it was an extensive job to cull them over and bring together the most important matters in orderly fashion. In spite of these obstacles, the job of expounding and recording policy has been carried on steadily; the material gathered together being arranged in the Company Policy Document which is appended. Side by side with the developing policy statement, the Managing Director is issuing, whenever necessary, an organized series of Company Standing Orders which set out established rules for applying company policy in the factories. In doing this, management is carrying out a role which would be commonly agreed as necessary—that of developing and maintaining consistency of policy and practice within the concern. But a further task which is not such a common practice has also been undertaken—that of drawing up and publishing a written statement in which policy is made as explicit as possible. Because it is written, such a statement can more readily be scrutinized, criticized, and changed in the light of experience.

Underlying this is the management's aspiration to make the firm as flexible and adaptive as possible. In this they base themselves on the view that communities which have a written and recorded history are more readily adaptive than those which have not. To record policy is to make an explicit digest of the most significant conclusions from past experience to assist in producing smoother and more effective change as future circumstances may demand. By creating a written record of the past and a formal statement of acceptable practices, the organization is freer to move away from its past and to try out new and informal practices.

But even the best of policy documents cannot dispose of all the underlying difficulties, for it is an impossible task to make policy completely explicit. There remains always a residue of unrecognized and unidentified aspects of the culture of the concern—such as the customs, the conventions and taboos, the unformulated criteria of status—which are difficult of access because they are deeply buried and often repressed. The identification and labelling of these is a never-ending process. However, the operation of such unseen and unwritten codes, important as they may be, does not diminish the value of having as explicitly stated a policy as possible. In Glacier the possession of an explicit policy, even though incomplete, has already allowed recognition of inconsistencies and gaps to take place more readily. As a result, further extensions have been made in the frontiers of the company's policy.

Factory Policy

Within overall company policy, as laid down by the unanimous decision of the three bodies composing the policy-making network, the Works Council at the London factory has the power on its own to establish principles governing local practices. Under these arrangements, a central responsibility of elected representatives is to collect and collate significant problems of concern to their constituents so that general principles may be clarified and developed. The function of the management is to run the factory within principles laid down and agreed, and to raise matters of principle which are not clear and have them discussed at the Works Council in order to get sanction for executive action. The management, by virtue of its full-time administrative role, carries the burden of responsibility for maintaining effective relations between the consultative and executive systems.

It is for the management to ensure that it has a sufficiently clear policy within which to work, and to make proposals when it has not. It is also a duty of management to collect the necessary material and information to enable the Works Council to arrive at well-informed decisions. This point was overlooked in the early stages of the standing committees. For example, the General Purposes Committee carried out an exhaustive factual survey of the time and cost of joint consultation before bringing back recommendations on how the problem might be solved. Now, when the Works Council raises a problem of this kind, the procedure is to delegate the matter to a standing committee, which may request the General Manager to arrange for appropriate executives to pull together the necessary information and to make tentative proposals which the standing committee receives and examines. It then makes its own recommendations to the Works Council, possibly in the light of discussion with other members of the factory, or with persons outside the factory co-opted for the purpose.

Confusion about Social Policy

One of the main difficulties still to be resolved concerns the diffidence, based on suspicion, of delegating to the management the responsibility for investigating problems of social relationships, and for drawing up proposals for principles to govern them. Responsibility for these matters has not yet clearly or consistently been assigned to any particular executive role. It is still left to the Works Council in many instances to take care of the difficulties. As a result, a feeling grew up in the factory that a split existed between social and technical policy; social policy being regarded as having to do with consultation and technical policy with management. The management, for example, were reluctant to regard the structure of joint consultation as a matter on which they were entitled to make recommendations other than as members of the Works Council. To do so in a managerial role would have been regarded as trespassing. No manager had the right to interfere in matters which affected the democratic rights and mechanisms of members of the factory. The present trend, however, is to recognize that joint consultation, along with all other matters of social policy, is as much a responsibility of the management as anything else, and consideration is being given to mechanisms for delegating responsibility for this function. The point of confusion has been that which lies between responsibility, on the one hand, for deciding policy, and cn the other, for planning it and putting it into execution. Clarification of the Works Council's responsibility with regard to the former, has allowed responsibility for the latter to be delegated to the management in all matters, including social policy, so that the elected representatives may not be overburdened with carrying out extensive and difficult investigations. The assignment of greater responsibility to the consultative structure, together with the sanctioning of greater authority for it, has in turn made greater demands for efficient and powerful management than have been experienced in the factory before.

CHAPTER ELEVEN

THE EXECUTIVE SYSTEM

The Operation of Authority

IN this chapter we shall turn our attention from system and role sanctions to the personal sanctions for the authority of the individual at work. Discussion will be concentrated on the command aspect of authority, which allows managers to give instructions to others in a subordinate position, as specified in the social structure. This emphasis arises from the fact that the most detailed and extensive case material so far obtained is concerned with command problems as seen in face-to-face, leader-subordinate groups. We shall consider the influence of those unconscious and irrational elements which, in affecting behaviour, disturb executive work, and also executive procedures for tackling the unconscious factors and for maintaining group co-operation in the executive system. Relations between groups in the executive system as a whole will be examined only so far as may be necessary to analyse behaviour in smaller groups. The management-worker split will be considered from the special and limited point of view of its use as a mechanism allowing face-to-face problems to be avoided.

THE ORGANIZATION OF AUTHORITY

The *executive system* is that part of the factory organization by means of which responsibility and authority related to the day-to-day production task are assigned to different persons. It is organized in a hierarchy of *executive roles*. By this definition everyone—the managing director, departmental managers, supervisors, specialists, operatives, and clerks alike—is in his executive role when carrying out his normal working duties. A *manager* is a person who occupies that type of executive role which carries responsibility for subordinates and authority over them. The differentiation between what are variously referred to as executive and functional managers, or line and staff managers and between executive and functional authority will be discarded. Instead we shall refer to two main types of role relationship within the executive system: *line relationships*—vertical relations between persons occupying superior and subordinate roles in the same direct line of command; and *functional relationships*—horizontal or diagonal relations between persons in their executive roles, who, however, are not in the same line of command. A line relationship occurs when a superior instructs or otherwise works with his subordinates. It is in this type of relationship alone that executive authority operates. A functional relationship occurs when one person is authorized by his own superior to seek a service or prescription from someone outside his own line of command who on his side has been made responsible by his superior for providing this service or advice. The policy at Glacier governing executive responsibilities and line and functional relationships is detailed in Appendix II.

Organizational units made up of leaders and subordinates will be referred to as *command groups*. We shall distinguish between primary command groups made up of a leader and his immediate subordinates at the next level of the executive hierarchy, to be called the *nuclear command*, and the larger unit made up of the leader and his total span of control, to be called the *extended command*. In this sense, the divisional managers constitute the nuclear command of the General Manager, and the London factories his extended command. The characteristic structure possessed by the nuclear command, namely, that of one person in charge of a number of people, will be referred to as *T-group structure*, a name chosen to describe the fact that on organizational charts such group structures resemble an inverted T. In the T-group one person is in charge of a number of people of varying age, all under contract to the larger organization and allocated to his command, for whom he is responsible only during working hours.

Chains and Lines in the Executive Hierarchy

The executive system is made up of chains of T-groups and lines of individuals. The form of the T-group chain is determined by the three basic types of T-group which occur in relation to position in the executive hierarchy. At the top is the Managing Director who, while having a nuclear command, is not himself a member of such a group, receiving instead periodic directives from the Board. At the bottom are the so-called primary working groups, operatives and office staff, under the command of a supervisor but in charge not of people but of factory and office machines and equipment. Between are the mid-level nuclear commands, whose personnel are both members of the nuclear command of a superior, and leaders of a nuclear command of their own. The Managing Director's superior is a group; he has no colleagues, but a group of subordinates. Mid-level management have superiors, colleagues, and subordinates. Members of primary working groups have superiors and colleagues, but no subordinates.

Communication in executive lines is through individuals. Each nuclear command leader is a gatekeeper for the members of his extended command. A breakdown in relations between any single superior and a subordinate means a failure of their particular part of the executive system to communicate with the rest. The appeals system is the formal mechanism that allows a subordinate officially to get past his own immediate superior. No individual can therefore act as a complete block in the executive system. There are also many informal and unofficial mechanisms for overcoming such blocks as result from superior-subordinate relationships, and we shall deal with some of these below.

SOURCES OF STRESS IN T-GROUP LEADERSHIP

The T-group as a social unit carries stresses peculiar to itself which arise from anxiety about authority in a situation in which one person is in charge of a group. In T-group structures, the leader is isolated when faced by his subordinates. He is also both dominant over them and dependent upon them. We shall now consider certain consequences of these features in relation to anxiety about authority.

Anxiety about Executive Leadership
Some of the aspects of the role of executive leader, as experienced by Glacier managers, which were sufficiently unpleasant to be avoided if at all possible, can be seen in the conflict

current among them between giving orders and extending democratic participation. This conflict was evident in the special connotations given to the words "democratic" and "authoritarian". Democratic behaviour, which was given high positive value, was equated with consultation. Consultation could refer either to discussions between a manager and elected representatives at any level in the organization (often by-passing one or more levels in the executive chain), or, alternatively, to discussions between an executive and his immediate' subordinates in which no authority is exercised—the subordinates being allowed to do whatever they might decide, so long as the matter had been fully thrashed out. Authoritarian behaviour, which was held in low regard, was identified with ordering anyone to do anything. Any command relationship within an executive line, in which a superior gave orders to his subordinates, even to implement general principles already agreed in consultation, came to be regarded, by many, as undemocratic and bad.

The constructive elements of these attitudes will be readily appreciated. Higher management was striving to eliminate practices by which some people gave orders to others without taking the views and feelings of those others into account. This they considered contrary both to social conscience and to practical efficiency. Their troubles arose because they had confused sanctioned authority with unsanctioned autocracy, and in so doing had undermined the processes necessary for executive work. If we take into account the anxiety about having responsibility and authority, the confusion between authority and autocracy can be seen as one method of rationalizing away the need for having authority at all. But many other methods also were used for avoiding authority, which give some idea of the discomfort produced by being required to carry responsibility and authority in an executive system. Some light may be thrown on the nature of this anxiety about authority if we consider the interplay of dominance and dependence, and the effects of a sense of isolation.

Dominance and Dependence

Having authority provides the T-group leader with an outlet for expressing his own power. This arouses whatever internal conflicts may surround his sense of power, such as conflict between unconscious doubts about personal capacities and ideas

276

of omnipotence. The co-existence of such opposites tends in practice to make it difficult to develop a balanced attitude towards using authority. Having a position of dominance stirs latent wishes to dominate for the sake of power itself as well as for the requirements of the job. But, to the extent that dominance is used for personal gratification, a profound sense of guilt ensues. Conflicts are engendered inside the individual himself, which may be increased if he feels himself to be in conflict with the accepted patterns of British culture. In fact, it often seemed to many of the managers that it was better not to give orders at all than to have to go through the acute personal disturbances associated with these conflicts. In these circumstances times of crisis are welcome. In a crisis everyone wants to be ordered about; the feelings of individuals do not count; they can be subordinated to the demands of the situation. Personal dominance becomes a socially sanctioned asset that can be expressed without guilt. Ordinarily, however, these sanctions do not hold. The manager is left to face the problem of controlling his feelings of omnipotence and of finding constructive outlets for his aggressive dominance.

As regards doubts whether he possesses power equal to his task, whatever doubts a manager may have in any case are reinforced by the anxiety of having a group of subordinates dependent upon him. The feeling that his behaviour is under their constant scrutiny may be exceedingly discomforting. How far do they judge his performance to match the demands of the situation? Moreover, a manager's subordinates are not only dependent upon him; he is also very much in their hands. How well he is able to discharge his responsibilities depends on how well they carry out his wishes. They can attack him by the simple technique of making mistakes and having things go wrong. To live up to his job, a manager has not only to carry out his own duties; he has to be capable of keeping under control both his anxieties that his subordinates may lack confidence in him and his impulses to chastise them should they make mistakes which might seem to him to have been intended to intimidate him. A paradox of executive work is constituted by the fact that the higher the executive, the greater is the number of people dependent upon him, but the greater also is his dependence: for the carrying out of his wishes is in the hands of an increasing number of people.

Analysis of Change

In so far as a manager's reaction—both to anxieties about inadequacy and to worry at the threat of being constantly open to attack—is to return with an even greater dominance and show of power, a circular process results. Such reactive dominance leads to stronger feelings of guilt; guilt arouses a greater desire to put things right, and hence to be in accord with the wishes of subordinates; submission increases both the feelings of inadequacy and the fears of attack, which stir a still greater impulse to dominate. Such a circular process leads either to paralysis by intolerable feelings of guilt, or to explosive reactions to excessive dominance; end-points which may be avoided by constructive executive skill and ability.

The Isolation of the T-Group Leader

To keep his own unconscious feelings under control becomes more difficult when a manager is face to face with his immediate subordinates. A sense of isolation results. Seen in its most extreme form in the position of the top executive, this isolation is found to affect every manager when faced by his own nuclear command. It appears a case of one against the many, the meaning of "the many" being multiplied by the different kinds of relationship for which the manager is responsible—those among his subordinates, between his subordinates and the next line of management, and between his subordinates and himself. Faced with all these, he has to contend with a welter of feelings, conflicts, and shortcomings which at times may seem overwhelming. Because he must act on his own, he has to some extent to have within himself confidence that he is powerful enough for his job. Such confidence presupposes an absence of the fear of dominating others and an absence of intense wishes to dominate them; a capacity to criticize without haranguing; a freedom from overwhelming feelings of guilt; a toleration of personal mistakes as well as mistakes made by others; and internal security and independence, and the ability to give security and grant independence to others. These, and other attributes that might be added, can be met only by the possession of genuine psychological maturity on the part of the individual.

The carrying of multiple roles may add to the problem of isolation. A manager who has a number of roles is responsible for a number of different groups and is isolated in each of them. To carry out these different responsibilities he must be able to make the appropriate shift of attitude each time he moves from one role to another. The sense of being split tends to result in increased anxiety about isolation, worsening the experience of having to move from one position of isolation to another. Assistance in facing these difficulties may be given by the factory as a whole to the extent by which it sanctions authority as a potential force for constructive good. It helps also when the manager himself is a member of a nuclear command in which he derives support from his own superiors and co-operation from his colleagues. But the only basic solution is in the manager's own confidence in himself and in his personal capacity to sanction his own authority; and it is in this regard that executive leadership requires personal qualities of security and psychological maturity.

Mechanisms for Getting Away from Responsibility and Authority

Having multiple roles in the organization allowed managers to evade direct executive responsibilities by taking up the most convenient role—that farthest removed from authority over the immediately present subordinate or subordinates. An example was the Managing Director's unconscious failure to perceive his role as a general manager. He would avoid his instruction-giving responsibility for his immediate subordinates (the divisional managers) and would adopt instead a form of consultative behaviour that was rather detached from the actual leadership of his divisional heads and more appropriate to his role as managing director. The effect was to cause a break at the top of the executive system; there being no direct line from managing director to division.

Another technique for avoiding executive leadership was to misuse the process of formal joint consultation. In itself practised for basically constructive purposes, joint consultation often provided a readily accessible route of escape from accepting responsibility for immediate subordinates by making possible easy and direct contact between higher management and workers' representatives. The consultative machinery came to be used as an executive by-passing mechanism. As a result, higher management was unable to assess attitudes at the junior end of the executive hierarchy. By comparison, its perception of the views of the elected workers' representatives was clear and immediate.

Reinforcing the obscurity of the functions of executive and

consultative activities, was the amount of philosophical discussion that took place about joint consultation. This was intended to clarify managerial policy, and in part did so, but in many respects was far removed from the real situation. It served the purpose of personal gratification for those taking part. To the extent that such purely personal needs were served, it made it simple for the joint consultative policy to be discredited by others as a "talking shop". This further contributed to the weakening of the executive, since discussion of, and training in, executive leadership were identified as being of a piece with the policy of joint consultation. They could thus readily be got out of the way. Impractical philosophizing was adversely compared with the real technological job of running an engineering factory.

Yet another technique was to bury oneself in amongst one's subordinates, an example of which was the General Manager relinquishing for a time the chairmanship of the Divisional Managers Meeting and becoming "just an ordinary member of the group, with no more say than anyone else", a form of behaviour which was practised by many other executives besides. Other methods were taking over the jobs of one's subordinates, and avoidance of the leadership aspect of the executive role, as reflected in the "delegating" behaviour of the General Manager early on in the study. It was particularly easy in the case of the higher executives to lose sight of their essential and direct responsibilities for their nuclear command by concentrating on their more general responsibilities for running their extended command.

EQUALS IN THE SAME COMMAND

Subordinates in the same nuclear command are related to each other by the common work task; they operate within the same immediate policy, and each carries a part of the overall task of the leader in such a way that the whole is integrated both by his activities and theirs. Teamwork may be regarded as the process through which a nuclear command maintains the wholeness of its executive task: by means of co-operative working relations among all members. Teamwork does not depend solely on leadership, but stems in large measure from the capacity of subordinates to work together. This is interfered with by the rivalries which frequently exist among them, and by the tendency to band together against the leader.

Relations among Equals

One of the strongest conventions in the Divisional Managers Meeting was that divisional managers should not criticize each other in front of the General Manager. When a difference occurred between divisional managers, the accepted convention was that the matter should first be taken up privately between the two. Failing a satisfactory solution, it would then be referred to the General Manager, the colleague having first been informed. If, as occasionally happened, the matter had been mentioned accidentally by one of those concerned while talking to the General Manager about other things, the latter would be asked to keep it to himself. On the few occasions that he failed to keep confidence it was regarded as a serious breach. These conventions, which will be recognized as common practice, served the purpose of maintaining the group's highly prized co-operative and friendly relations by keeping at as low a level as possible the open appearance of rivalry, jealousy, or hostility. They also served the less conscious and conflicting purposes of co-operation in preventing the General Manager from finding out about mistakes, and yet assisting him to find out about them at the same time through accidental remarks in casual conversation. The importance of these conventions is brought out by the fact that it was only when some of them were identified that the General Managers Meeting was able to proceed with the task of clarifying executive policy.

Among the superintendents, phenomena of rivalry were particularly noticeable during the time when they neither were given clear-cut executive leadership nor held a defined position in the executive structure. Anxiety about personal prestige and the status of their roles was at a height; it being uncertain whether technical, administrative, or practical skills were most valued. They were angry with the higher management, but were conspicuously unable to express this in a way that would secure them the action they desired. This was in part because their rivalry and uncertainty *vis-à-vis* one another made clarification of the situation a potential threat, in that clarity might have led to higher status being given to one or other of their roles; and in part it was because, by not taking up their difficulties, they were co-operating in an indirect attack on their superiors, for whom they were making things more difficult.

How subordinates were to co-operate in making their feelings known to their leaders was a problem encountered in many spheres. The Divisional Managers Meeting was strongly influenced by the idiosyncrasies and special interests of the General Manager. How could his subordinates best range themselves around him so as to keep his enthusiasm within practical limits and his interests within profitable channels, yet themselves avoid acting as a brake on reasonable progress? The workers in the Service Department, in the early stages of the discussions, were anxious over changing to the flat-rate system of payment which most of them desired, fearing they would be driven by their superiors; they did not know what to do with their own aggressiveness against them, or how to co-operate in achieving their goal.

Status Rivalry, and Co-operation Against the Leader

Subordinates have the task of getting on with each other, and getting on with their leader, separately and together. But who is to get on best with the leader? What happens when there is disagreement with him? In the group of equals, whose position really has the highest status, and whose the lowest? Let the subordinates band together, and could they not do their leader's job and do it better, and so get rid of him? But if they get rid of him, would they not fight among themselves? Questions such as these are a paraphrase of the attitudes which unconsciously affected relations in nuclear commands. Many of these difficulties arose from the nature of the sub-structures existing in a nuclear command: the leader himself, the subordinates as a group, a series of pairs composed of the leader and each subordinate, and the total group of the leader and the subordinates taken together (not to mention sub-groupings of an intermediate size). In a nuclear command the leader must accept the right of his subordinates to co-operate with each other independently of him. They must accept his right to act independently of them. Subordinates must also accept each others' rights to their own jobs and status, and an independent relation with the leader.

When the leader meets separately with one or other of his subordinates, he forms pair relationships. These pairs are perceived by the others as destructive of the group as a whole, an observation consistent with experience in groups of other kinds. If the leader is with one subordinate, he has withdrawn from the rest; feelings of jealousy and rivalry are aroused because the favoured subordinate is felt for the moment to have the position of highest status. This is not to say that phenomena of this kind are conscious and rationally phrased by individuals in this way. Primitive emotions of this kind are aroused deeply inside the individual when he is a member of a T-group, and effect his attitudes and behaviour in a manner which he cannot easily control, since most of the underlying feelings are not usually accessible to him. Rivalry for the attention of the leader is intensified by a desire to be the leader. Along this route lies one means of solving the whole problem, for to gain "complete control" over the situation would be to regain "complete control" over oneself. Status anxiety and allied phenomena in the group have a personal foundation in the internal worlds of the individual members. What they try to control in the group reflects what they are attempting to control in themselves. Jockeying for position, and status rivalry, represent attempts to gain a sense of personal stability by getting a secure and definite position in the social structure.

In the situation of the leader *vis-à-vis* all the subordinates together, there is another source of disturbance. Banded together, subordinates may feel more powerful than their leader, but they were rarely observed to use their power, even to the extent only of open criticism. Inability to criticize the T-group leaders may be explained by the interaction of at least three factors: the fear, often explicitly stated, that the others could not be counted upon; the feeling of guilt which was stirred by the intensity of the unconscious desire not only to criticize the authorized leader but completely to eliminate him; and the anxiety over losing the guidance and protection of the person upon whom the subordinates were dependent. So frustrated, the desire to attack rebounds on the group, sharpening the internal rivalry situation.

UNCONSCIOUS MECHANISMS FOR HANDLING STRESS

Some of the difficulties in executive work which were observed during the project have been related to the involuntary operation of feelings of rivalry, aggressiveness, domination, dependence, and isolation, and the automatic reactions of fear and guilt, which occur in those who occupy positions in an authority system structured in the form of T-group nuclear commands. A number of unconscious and automatic mechanisms operated to offset these stresses. Examples of such mechanisms are presented below.

Scapegoating

One general technique was for leader and subordinates alike to avoid their own immediate tensions by attacking some other group. For such a purpose the split between the specialist and the production departments was ready to hand, allowing one to blame the other. Mid-level executives could take out on their own subordinates their troubles with their own superiors, and ascribe to those below their own desires to band together against those above. In the primary working groups, those above could be made the butt of displaced anger. But so also could elected representatives, and the district union officials, who on occasion were held to be the cause of difficulty and made the target for hostility.

Because scapegoating served the useful purpose of providing an outlet for pent-up aggression, any steps which could be interpreted as doing away with a convenient scapegoat were resisted. Moreover, the actions of those who tried to introduce these steps were resented. Attempts to specify policy governing functional relations met with resistance because a clearer policy would have meant less scope for mutual scapegoating. Rational arguments by some workers that the management seemed fair were rejected by others; and equally rational arguments by some managers that workers could be trusted to behave responsibly were resisted by their colleagues. The elimination of scapegoating requires resolution of the stresses giving rise to the need to find outlets for hostility.

Willing Acceptance and Active Resistance

Getting rid of stress on to someone else was not the only way that members could deal with stresses displaced downwards on to themselves. Individuals, in Glacier as elsewhere, found satisfaction in feeling that they were suffering at the hands of their leaders, or enduring pain and trouble in order to protect others. Unwitting collaboration occurred between superiors and subordinates to maintain a position in which the subordinates could think of themselves as bearing up to stress. This was noticeable in the offices and in groups which provided services, a considerable number of whose members gained satisfaction from having constantly to right matters which had gone wrong. Attempts by others to make improvements so as to reduce bother and fuss were not always welcome, and even gave rise to anxiety.

Active steps were taken, particularly by the workers, to avoid being made the final repository for stresses displaced down the executive lines. The appeal channel, the consultative channel, and the trade union channel were used when the attitude of his immediate superior was such as to make a worker feel that it was no use trying to take his grievance up that way. The appeal channel was used almost exclusively by workers. The consultative channel also, as regards redress of grievances, was used mainly by the workers. The stated reasons for seeking redress were not always the main reasons, but the use of these mechanisms was a conscious and direct defence in comparison with the unconscious motivation involved in scapegoating, or the willing acceptance of stress.

THE ADAPTIVE EXECUTIVE PROCESS

So far as disturbing feelings and the cultural mechanisms for balancing them operated unconsciously, they were not easily amenable to control. They would have gone unchecked, causing profound difficulties, but for the existence of counteractive tendencies. The main force for getting these primitive and involuntary reactions under control, and their energy directed into constructive channels, was that of conscious executive planning and work, which will be referred to as the adaptive executive process.

Organizing the Executive Task

In an hierarchical organization, an executive task which is delegated to a person has the characteristic of being both a part and a whole at the same time. It is a part in the sense of being one part of a larger whole, other parts of which are delegated to other people; but at the same time his own part has a wholeness for the individual executive as the entity for which he carries total responsibility, and may delegate further as he sees fit. Only members of groups at the end of an executive chain are without opportunity for further delegation of their own work. In addition to the shop floor, these groups include technicians and specialists whose executive task is not to administer the work of others but to work on their own. Apart from these cases of people at the end of an executive chain, executive work consists of organizing the task so that the responsible executive keeps an appropriate part for himself, while delegating the rest to his subordinates. But once the task has been split up in this way, wholeness must be regained by integration, which introduces the need for a precise and practical co-ordination of working relations within a nuclear command. The assignment binds people together because each person holds a part of the job; but performance can issue as an operational whole only so far as the social structure, customs, and working relationships in the group concerned will allow. How the manager keeps his task differentiated but at the same time integrated within his nuclear command determines the effectiveness of his work. We shall consider two methods for accomplishing this two-way task: making local policy or task-policy, and face-to-face meetings of nuclear commands.

The more a task is broken up in the process of delegation, the more essential becomes articulation of a task-policy or general statement regarding the relationships between the parts. The designing of task-policy is the primary mechanism by means of which a nuclear command maintains the wholeness of its work task in the face of the need for delegation. A properly conceived and clearly stated task-policy allows each person to work within the same general framework as everyone else, and perceive his own part in relation to the rest. Task-policy has many different aspects. One is job layout, for which shop management is responsible with technical advice from the Production Engineers. Another is the culture adopted by the members of a nuclear command to regulate the way they are to work together, help, and criticize each other in sorting out technical and group problems; this is a matter which has much local variation within the framework of the policies of the larger organization, and which ordinarily is rather more implicit in the ways of the group than explicitly comprehended or stated.

How the T-group leader forms his task-policy is a matter of considerable importance for group relationships in his command. On some aspects he will receive advice from outside, as in the case of job layouts or piece rates. But it is just becoming accepted at Glacier that such outside advice can only be a recommendation, and that it is entirely up to the manager concerned whether he follows it. So long as he is held responsible for the task assigned to him he must be given the authority to carry it out in his own way, within the general framework of factory policy. It is only on this condition that a manager is in a position to take on the responsibility of building good teamwork, whether he proceeds by setting his local policy by himself, or reaches it through two-way discussion with his subordinates. Where he himself sets task-policy, how the task is carried out will depend on how far his subordinates are willing to follow him; where there is a greater degree of discussion, there is greater possibility for the members of the group to sanction each other. The breadth of sanction required will depend on the nature of the task; it being part of good leadership to know when to bring the total group together.

Building Group Cohesion

A well-done job, carried out within a sound policy which allows all the members of the group to perceive the structure of their task relations, is the basic requirement for overcoming the feelings of fear and guilt which confound T-group functioning. The way the task is carried out allows the members of the group to test their personal anxieties against the realities of the work situation. A piece of work well done in co-operation with others has the reassuring effect of proving anxiety unnecessary. A good task means a good group, so that the control of disturbing feelings becomes more secure and the expression more free of both complimentary and critical attitudes. It is this effect of damping anxiety about disruptive forces that goes far to explain the characteristic feeling of satisfaction which accompanies the completion of a work task in constructive co-operation with others. The importance of regular opportunities for meetings of all members of a nuclear command may readily be seen as one means of supporting sound policy-making and reality-testing on the job. Bringing everyone together makes it possible for the task of the group as a whole to be experienced in a first hand way, and for task-policy to be argued and clarified. In discussion, leadership can revolve from person to person in the group according to skill and knowledge. Everyone therefore may have the same chance of experiencing responsibility for the group and to the group as the result of leads which he himself has given. At the same time the formal leader retains his position of authority. In the process of participating in making task-policy, members of the group unconsciously match their disruptive feelings against the realities of their relationships as they actually experience them in the group. This type of reality-testing reinforces the realities of their experiences with each other in the course of carrying out the job. These mechanisms of reality-testing in nuclear command meetings are a means of assisting the realities of good work to cope with the more involuntary anxieties aroused in T-group relationships. The resulting social climate is one that more easily permits sound day-to-day executive leadership to proceed.

Just as a job well done produces a sense of group satisfaction, job failure produces depression and anxiety. When things go wrong, the dependence-anxiety, aggressiveness, fears of attack, rivalry, and guilt are all stirred up. The way in which mistakes are dealt with is the real testing ground for the quality of executive leadership and the soundness of follower relationships in the T-group. Intellectually the regular scrutiny of errors of judgment, mix-ups, inefficiencies, and incorrect actions of all kinds, and sincere attempts to understand the causes of these troubles, serve much more than the obvious purpose of helping the group to learn and to plan how to avoid similar happenings in the future. Eventually they have the profoundly reassuring effect of proving that the group is not falling apart, and that the antagonisms of members to each other are not of the kind from which disastrous consequences follow. Such proof is only obtainable where leader and members alike can demonstrate to each other that they can criticize and tolerate each other's mistakes, while refusing to countenance continuing inefficiency. Where mistakes and concealed inefficiency are ignored, surface calm may reign, but failure and disruption ensue, since the involuntary group fears are left free to work their havoc. The leader who cannot give constructive criticism to his subordinates, both as individuals and as a group, is in no position to lead, and the group which is too anxious to note and raise objections to his actions is in no position to follow him or to sanction his leadership.

Maintaining an Effective Nuclear Command

The manager needs to be supported by accurate information from his nuclear command, by continuous training on the job to meet changing requirements, and by sound organization and administration. The need to establish a process of *two-way*

288

communication is related to the fact-finding function, or that of getting accurate information from subordinates so that the best decisions can be made. Every manager is in the hands of his immediate subordinates when it comes to getting regular information about the effects of the instructions he issues, and about difficulties with which he should be dealing in his extended command. In the absence of efficient fact-finding, troubles pile up and are left to be discovered at a late stage when they are less easily remedied. In practice, successful fact-finding requires that the relation between a manager and his nuclear command shall be such as to allow free talk on all matters, whether to do with technical problems, relations within the nuclear or extended command, factory policy, or the attitudes, behaviour, and policies of the manager himself.

The relationship between executive and consultative work was much less confused once the fact-finding function was perceived as part of executive work. The term "consultation" is now becoming reserved exclusively for those cases where elected representatives meet as such either with each other or with the managers concerned. The process of formal representation has been distinguished from that of two-way communication including fact-finding, within the executive system. The importance of this verbal change is that it marks the end of irresolvable arguments as to whether "two-way consultation" was necessary in the executive system when formal joint consultative procedures existed. Two-way contacts of some kind between managers and their subordinates are inevitable, not merely desirable. One of the arts of management is to arrange these contacts to allow open two-way communication of the frankest and most comprehensive kind to take place so that the most effective executive work can be done.

The *training function* is the expression of the manager's responsibility for ensuring that his own subordinates are sufficiently skilled to carry out their tasks. Placing this responsibility squarely on the manager himself was a marked change at Glacier, for primary responsibility for training had previously been laid at the door of the training department. With this shift in responsibility back to the manager concerned, the training department is able to return to its prescribed function of providing training services, facilities, and advice, but without having to carry the burden of deciding whether people are adequately trained for their jobs, or what they should be trained for. With the primary training responsibilities for his subordinates laid upon the manager himself, there was thrown into bold relief the fact that his ordinary behaviour towards his subordinates was the main factor in training. A manager's behaviour towards his own subordinates, while not necessarily identically patterned on the behaviour of his own superior towards himself, is nevertheless markedly influenced by it. The actual behaviour of the top executive is therefore the most powerful training influence operating in the total organization, and special training programmes conducted by the training department, which were too far removed from the real executive behaviour patterns in the organization, were regarded as academic, no matter how commendable they might be on grounds of social philosophy.

To obtain sound *organization and administration*, personnel must be deployed within both nuclear and extended commands according to a known and understood organizational structure. Yet to get an organizational structure according to which people could actually and consistently behave as though they were in charge of those of whom they were supposed to be in charge, proved difficult. The problems involved were thrown up sharply when the practice of managers meeting regularly with their nuclear commands spread through the executive system. Stress occurred at those points where a manager's subordinates, as shown on the organizational chart, did not tally with the actual subordinates through whom he ordinarily worked. Changes in organizational structure are taking place where such stresses have been located. Placing the emphasis on each manager's direction and control of his own nuclear command has been a central factor in teasing out the above executive responsibilities. With such an emphasis, a clearer and sharper definition has been given to relations within executive lines. Out of this a criterion for the adequacy of the firm's organizational structure can be derived, a sound organizational structure being one in which each manager can carry out his duties by working only through his immediate subordinates and with his own superior and colleagues.

Sanctioning the Authority of the Individual Executive

In summary, the adaptive executive processes which we have now considered are the means whereby the manager receives personal sanction for his authority from his superiors, his colleagues, his subordinates, and from himself. A person must first have sanction from his own superior, in the sense of his superior having confidence in him. Except in cases where a superior keeps the members of his nuclear command entirely separate from each other in their work, some degree of co-ordination of effort among colleagues is inevitable; to this extent colleagues must sanction each other's authority in the sense of a mutual acceptance of each other's roles and responsibilities. The importance of this sanction can be illustrated by the disruptive effects which occur when, through rivalry, colleagues no longer sanction each other's authority. Sanctioning of a manager's authority by his subordinates is equally necessary, because the position of authority not only gives the right of command, but also makes the person giving the command temporarily dependent on his subordinates for the manner in which they carry out his commands. The superior-subordinate relationship is a combination of mutual dependence and mutual dominance, with the appeal channel providing the check and balance against unbridled dominance from above. But no amount of sanctioning from outside will be of any avail if the executive, by reason of his personality make-up, is averse to accepting authority. Feelings of anxiety and guilt about omnipotent and autocratic dominance, and fears of not measuring up to the task, prevent secure personal acceptance of authority, and cripple executive work. Final personal sanction comes from the ability of a manager to be sufficiently secure within himself to be able to tolerate either a group of subordinates who are powerful, and capable therefore of applying strong sanctions against him—indeed, only from such a group can strong positive sanctions also be obtained—or a group of subordinates who, because they are not particularly able, cannot sanction their leader powerfully and, hence, require him to draw upon strong sanctions from within himself.

THE EXTENDED EXECUTIVE CHAIN

Until recently, the problem of leadership was widely conceived at Glacier, as it is in industry generally, as being a disturbing feature at the level of the section supervisor. Because section supervisors were directly responsible for workers, extensive leadership training courses were initiated for them. As regards those higher up the executive system, it was assumed that they must necessarily possess leadership skills to have got there; if not, that leadership skills were not needed so much because subordinates at the higher levels could do without much direct leadership. This attitude has become completely reversed. It is now accepted that the pattern of leadership adopted by higher management sets the pattern for the rest. The chief concentration of top management has been turned towards getting its own executive behaviour and policy clarified. This process has begun to move down the executive system, but leaving largely untouched as yet the problem of the section supervisor.

The Top Management Pace Setting Effect

The widespread pattern of avoiding direct executive responsibility for immediate subordinates demonstrated how the behaviour of top management sets the general tone of behaviour. Recently, as the top management approach changed, with the Divisional Managers Meeting becoming the General Managers Meeting, those divisional managers who were not already doing so have commenced meeting regularly with their nuclear commands. This pattern is now spreading downwards through the organization. One of the more important factors leading to this change was the working-through of relationships within the top management group. The quality of their executive work stemmed from the degree of insight they had into the nature of the forces affecting their relationships with each other. Since the top management have recognized that in order to maintain a telling leadership role they must straighten out their own relationships first, a new set of conventions is being elaborated by them. Among these is the pattern of the General Manager moving no faster than is consistent with a coherent and integrated working relationship with his subordinates. This demands that the divisional managers should be capable of putting aside differences whenever a situation calls for rapid change and progress.

Before the nuclear command problem was tackled, managers saw themselves, and were seen by others, as the leaders of the other managers in their extended commands and having most contact with those closest to them in the hierarchy. But they were not regarded as the executive leaders of the bottom stratum of workers. With the realization of the special character of the relation between a manager and his nuclear command, a more differentiated picture of his total command was obtained, and two questions were thrown into sharp relief. The first was that of his responsibility for the relationships between individuals in the executive lines, and between T-groups in the executive chains radiating out from him; this will be considered in the section on communications in the following chapter. The second was that mentioned above—the question of his responsibility for giving leadership to all in his extended command, including those at shop floor level—a vexed question which has political as well as administrative implications, and to which we shall now turn our attention.

The Break at the Bottom of the Executive Chains

The straightening out of relations in higher and middle management having been undertaken, the problem of the relationship at shop floor level between the section supervisor and his team of workers has re-emerged with new force and clarity. The section supervisor had been scapegoated by both higher management and workers' representatives to avoid taking account of their own difficulties. But now, more reality could be introduced into the perception of the shop floor problem. It became apparent that the notion was widely held among managers and workers that the executive system ended at the section supervisor. His relationship with his workers was regarded as somehow different from his own superior's relationship to him. The view was variously expressed as the belief that workers should take up all policy matters through their representatives, or that the section supervisor should only instruct his workers and not have to discuss his instructions with them.

Glacier section supervisors with a team of ten to twenty workers have become an identifiable managerial group only in the past five years. They derived from the setters (operators who could set as well as operate a number of machines) who were selected for technical skill rather than for leadership qualities. Although both leadership and technical skills are now required, the earlier picture of the section supervisor as a setter persists. There has been a fear that if managers come to be regarded as on-the-job leaders of the workers, this might put them in a position to interfere with the rights of the trade unions. This resulted in the idea that the supervisor may tell his operatives what work he wants done, and see that the work gets done. But it is the shop steward or representative who has both the responsibility and the sole right to transmit to higher management the views and demands of the workers about their work—and he, by tradition, was concerned mainly, if not exclusively, with wages.

Recently, however, the belief has grown that for the executive system to work coherently, it must run from the Managing Director to, and include, the workers, at shop floor level. The section supervisor must be more than a groupless leader. He must have the same leadership, fact-finding, training, and organization responsibilities for his nuclear command (his team of workers), as has the General Manager for his divisional managers. Any differences are not differences in principle, but differences in practice, related to the differences in the problems to be dealt with. The worker, just as the manager, is in his executive role when carrying out his particular work task. And in that role he requires a two-way executive relationship with his superior. True, in his role as a member of the stratum of hourly-paid workers, he has access to the trade union machinery or joint consultative system through his shop steward or other elected representative. But he is no different in this respect from any manager who, as a member of a staff grade, also has access to the consultative machinery through his own elected representative and his staff committee. The link between the section supervisor and his team of operatives in their executive roles is not only an integral part of each executive chain. It is that special link which finally brings higher. management into contact with the production process as experienced by the worker who does the job. Any weakness at this point impairs management's capacity to manage, and militates against satisfaction on the job for workers.

The main executive problem now facing the factory is how to get a set-up in which day-to-day matters having to do with the implementation of policy as experienced on the job—for example, methods of payment, effects of new technological methods, reasons for scrap, rate-fixing practices and results, reasonableness of production schedules, quality of machinery and tooling, fairness of discipline—will all be taken up in the first instance via the executive system. To take the case of a radical change in production method: under the consultative procedure, such a change, a matter of policy, would only be introduced after full discussion with workers' representatives. They would consult their district union officers before agreeing. Agreed policy would be implemented through the executive system, and at shop floor level by the section supervisor through his workers. The section supervisor, having implemented the policy, must deal with the problems of his subordinates arising out of it. Problems with which he cannot cope must be transmitted higher up his executive line for action. For their part, when they are dissatisfied with their supervisor, the workers have both the appeals channel and the consultative channel through which to take up grievances.

The Management-Worker Split

The perception of the executive system as extending right to the shop floor for all purposes may seem a slender return for much work. This interpretation would overlook the deep-seated resistance, both in the management and the workers, to interference with the management-worker split. The split can have only limited resolution inside one factory. It is the outcome of the general conflict in society between management and workers. The intensity or quiescence of this conflict nationally, sets the limits for industrial relations locally. But, even within the limits allowed by the national situation, there was resistance at Glacier to tackling the break in the executive chains at shop floor level. The split had come to serve other group needs as well. Management saw the demands of the shop floor working groups as potentially disrupting if workers were allowed too free expression through the executive system. With some relief they left to the workers' representatives the job of being responsible for bringing forward the demands of workers. Ordinary day-to-day complaints or demands could be sidestepped, to be faced only from time to time when feelings had heated to boiling point. In between, or so they wished to believe, the technical job of production was subject to a minimum of interference. On their side, the workers found it an advantage to have a management which remained split off from themselves. It gave them a scapegoat upon which to vent hostile or aggressive feelings, whether or not these feelings were tied up with the job. Having such an outlet meant that differences with other workers need not be squarely faced, as, for example, was recounted in the case of the Works Committee.

Because of these mutual advantages (and others which are still under investigation), there was unconscious collusion between the management and the workers to maintain the split in the executive system, and to have all demands of the workers ironed out through the consultative system. A fine balance was struck. Sufficient supervisor-worker contact about the job was maintained to keep the work going. How much contact was a matter of how good were the relations at any particular time. Seen in this light, the decision to tackle the problem of creating a coherent executive system means giving up this (from one point of view, useful) collusion. Three main factors are offsetting the anxieties both of managers and of workers at the prospect of no longer being able to stave off facing these problems: first, the growing recognition that problems which are not tackled directly are expressed indirectly and cause even more trouble and disruption; second, the experience that difficulties in group relationships can be successfully tackled and worked-through; and third, the realization that the effect of having a split in the executive system is to place higher management executively out of contact with production, and to make the workers, when on the job, into a leaderless group.

Under the now emerging policy, the section supervisor is charged both by his superiors and his workers with the responsibility for having first-hand knowledge about his team's work, problems, and demands, and for seeing that their views are transmitted upwards. The consultative system is left free for its agreed purposes—for getting redress of grievances when groups are not satisfied with the operation of the executive system, as well as the route for expressing the common interests of the hourly-paid workers and of the three grades of staff in making company policy, and so to cater for their special needs. In the case of the hourly-paid workers and other union-organized groups, the consultative channel also allows expression of their special interests as members of the organized working class outside the confines of the factory. How far the members of the firm will be able consciously to carry the process of distinguishing between their responsibilities in their executive roles, their status group and representative roles, and their trade union roles, and to act accordingly, has yet to be discovered. It will depend on how far shop stewards and other elected representatives are both willing and able to forego exclusive concern with details of day-to-day factory work. An additional ingredient is required —greater preoccupation with the soundness of policy in their constituency, and with the means by which the policy is being implemented. The extension of executive leadership to the shop floor, through making for greater efficiency in dealing with ordinary problems, places greater responsibility on elected representatives, by making greater demands upon them to adopt a broader frame of reference for their activities.

CHAPTER TWELVE

SOCIAL ADAPTATION

The Nature of Change Processes in the Factory

IN considering the pattern of social change in the factory, we shall examine the interplay of social structure and social change processes as seen in the problems of the role clarification and of establishing effective communication. The process of working-through will then be described, and the function of working-through examined in the light of its effect in increasing social adaptiveness. By way of summary, a general interpretation will be made of the present phase of development of the factory.

ROLE CLARIFICATION AND ORGANIZATIONAL CHANGE

Many of the problems surrounding role clarification were exemplified in the top management group. The Managing Director carried at least three roles—Chairman of the Board, Managing Director of the total organization, and General Manager of the London factories—his general managership having remained undifferentiated from his managing directorship. It may seem a simple matter to have clarified this particular issue. One may wonder why its significance had not been earlier perceived. To understand the reasons for confusion the emotions surrounding these different roles must be taken into account, and the fact that different roles require different kinds of behaviour as well as relationships with different groups of people.

Role Clarification within Higher Management

A re-examination of the roles of the General Manager brought out the kind of conflict that can be engendered when one individual occupies roles that are closely related to each other,

298

but entail relationships with different groups. His two roles of managing director and general manager were not distinct in the minds of others, and at Divisional Managers Meetings it required effort on his own part to keep them separate. In the role of managing director he was responsible for the co-ordination of the total company, acting through at least two subordinates —the General Manager of the Scottish factory, and the General Manager of the London factory (namely himself). His contact with the divisional managers of the London factory was therefore indirect, but in his role of general manager he was immediately responsible for the leadership of this group. It was in this role that he felt uncomfortable, for it was here that order-giving behaviour was required of him. So long as these two roles remained undifferentiated, it was easy for him to retreat to the managing directorship, and he discovered that he had rarely occupied his general managership in the sense of assuming the leadership of his divisional subordinates.

Another indication of the manner in which group relationships may diminish the accuracy with which roles are perceived was the way that the Divisional Managers Meeting oscillated between being a Board of Directors (taking group decisions affecting the whole company), an executive management committee (taking group decisions affecting only the factory), and an advisory meeting of an executive and his subordinates (with the General Manager taking the decisions, after suitable discussion). To straighten out this confusion required months of discussion during which the experiment of the General Manager relinquishing the chair took place. Under pressure of the practical necessity of stating the functions of the Divisional Managers Meeting in the policy document, it was ultimately decided to adopt the form of a co-ordinating meeting called by the General Manager in which he would be responsible for all decisions taken but would seek the advice of his subordinates as far as possible. To implement such an arrangement, even when the situation was straightened out on paper, still required acceptance by the General Manager of his duty as executive leader, and constructive follower-ship from the divisional managers. It is here that trouble comes, for people's behaviour does not automatically change in accordance with the legislated state of affairs, or indeed, with what they themselves may rationally believe to right.

Analysis of Change

Resistance to Role Clarification

Perception of individual roles was in fact a matter not simply of intellectual clarification, but involving the feelings and attitudes of the individuals in the roles. Individuals occupying roles have perforce to make relationships with other persons, as specified by the organizational structure, in a manner set by the culture of the organization. Occupying a role involves conforming to external social demands, and the more clearly the social structure is perceived, the more clearly the requirements of the situation are seen. For the Divisional Managers Meeting to perceive that the Managing Director was also the General Manager, created a situation in which the break in the line of authority between the General Manager and his divisional managers had to some extent to be put right.

A person who occupies a number of roles acquires relationships with a number of different sets of people, not to mention different relations with the same people. A role divided between two or more persons imposes on those concerned the necessity of co-operating as a role-carrying group. Changes in social structure entail changes in formally established relations and, hence, also require the working out of new personal networks. These potentially confusing and difficult inter-personal connections demonstrate the necessity of adhering rigorously to the role required by each particular situation. Accomplishment of such behaviour becomes possible only when the roles themselves have been precisely defined.

Role Confusion as Social Defence

The resistances encountered against perceiving roles and role relationships demanded special explanation in view of the tenacity with which inaccurate perceptions were adhered to. The explanation adopted in working-through these problems was to assume that role confusions were strongly, though unconsciously, motivated. Role confusion is an unconsciously motivated defence to which individuals have recourse in order to avoid the anxiety produced by disjunctions between their personalities and the demands of the roles they carry. If this picture of role confusion is correct, one must expect that the task of obtaining flexibility in organizational structure will be hampered by strong opposing forces. Successful adaptation presupposes in those concerned a genuine wish to change as well as illumination of the nature of their resistances. How much this was the case is demonstrated in the difficulties experienced in reconstituting the Works Council.

It could be said that there was a widespread desire in the factory to avoid the very notion of role and with it the complexities of role carrying. This was a vital issue for the Research Team because it was essential that its independent role should be perceived and understood in the firm. As it turned out, the tenacity of the Research Team at the very beginning of the project in working out its own role in co-operation with members of the Project Subcommittee was an influential factor in assisting the conception of role carrying to take root. Members of the Research Team were seen by members of the firm in different roles —as co-planners on the Project Subcommittee, and as social therapists in various groups. In each case great care was taken to specify the nature of the role being occupied at the time. As the project advanced to tackle some of the more disturbing problems, the security derived from precise knowledge of the role carried by the Research Team was appreciated, strengthening the motivation of increasing numbers in the firm to examine their own roles.

COMMUNICATION AND CHANGE PROCESSES

Communication includes not only verbal statements and instructions, but non-verbal and behavioural messages. It is the sum total of directly and indirectly, consciously and unconsciously, transmitted feelings, attitudes, and wishes. Communication is an integral part of the process of change, and occurs whenever social equilibrium is upset. Ordinary everyday events, like an order from a customer, cause a series of communications to take place that lead to the production and dispatch of so many bearings to a particular place. The piling up of economic stresses or the effects of other large-scale technical or social changes may lead to extensive and protracted communications involving large groups of people all at once, and requiring widespread changes in the organization, such as were brought about by the redundancy crisis and the introduction of the Courtenay plan.

The effectiveness of a communication system depends on the quality of the relationships between the people involved, a feature described in the findings from the communications studies carried out with the Works Council and the Divisional Managers Meeting. The effectiveness of the communications taking place was not judged on the extent to which there was a free flow of all communications, but rather on the quality of the selectivity in the transmission of information. Efficient communication was brought about by controlling relations within the organization in order to keep communication at a level that allowed changes of maximum use to be carried out, while damping less important matters. In other words, not all barriers to communication were taken as symptoms of trouble.

The creation of constructive barriers against too free communication will be referred to as *adaptive segregation*. By contrast, the erecting of barriers against communication, as a defence against stresses between groups, so that the transmission of important matters is obstructed, will be referred to as *maladaptive segmentation*.

The Process of Adaptive Segregation

The process of keeping communications between groups at an optimal level by the creative use of selective barriers can be seen in a number of regions in the factory. One example is the apparent dis-interest of employees in the Works Council. This attitude was looked upon as apathy by members of the Council, but can equally well be regarded as containing elements of creative segregation. Broadly speaking, the joint consultative bodies attract those people who are interested in organizational activities. Active participants who had also worked elsewhere had been equally active in their previous places of employment. This selectivity was intensified by the fact that consultative activities took place as an "extra", usually outside working hours, so that only those who derived personal satisfaction were attracted. The majority, at all levels, were content to confine their interest to their normal jobs and to their immediate working groups. While glad to know that the consultative machinery existed (should they have occasion to use it), nevertheless they did not encompass it as a regular part of their lives in the factory. Having got on with their normal jobs during the day, they wished to get home to other interests at night.

Viewed in this way, the complaint of elected representatives that their constituents were apathetic can be taken as meaning that the constituents were not interested in the same free-time activities as were the representatives. The joint consultative machinery provided an outlet for the special interests of the representatives, and a democratic mechanism for the factory as a whole. The absence of continued and intense participation by everyone did not spell failure, so long as an efficient organization was kept in existence which could operate when required.

In normal circumstances the main communications route between the Works Council and the factory was via the executive chain. The burden of discussion and preparation of policy fell within the executive orbit, with elected representatives agreeing or modifying what was proposed without extensive reference to constituents. In this factory the basis of satisfactory consultative work, paradoxically enough, was efficient two-way communication within the executive chains. Because the proceedings of the Council were public, factory members had a full measure of control over the extent to which they wished to have a direct say. When they wished to participate, as over the redundancy crisis, there was little difficulty in communication with representatives. When this wish was absent, pressure to take part was of little avail.

By the process of adaptive segregation, the elected representatives were charged with getting on with the job of consultation until such time as the constituents demanded to make their voices heard. The members of the factory could continue undisturbed in their normal work (and also their home lives) until crises arose demanding the attention of all, at which times the barrier was non-existent. Another form of the process was instanced in the Divisional Managers Meeting, where feelings about people were not transmitted in order to avoid arousing unnecessary anxiety. In the interests of peace of mind for the others, the convention stood that members did not disclose all their personal misgivings. These conscious methods of preserving maximum harmony in the group were over and above the blockages and failures in communication arising from the rather more unconscious causes, which, because they were unconscious, were not easily susceptible to control.

Barriers to communication are never clearly the result either of segregation or segmentation. They tend to be a mixture of both features. In the above examples segregation predominated, making possible both the setting aside of unimportant matters, and storing for future action important issues which could not

v   303

be dealt with immediately. An extreme example of adaptive segregation was the shutting off of technicians and research staff while engaged on the development of special ideas.

Segmentation as Maladaptive Defence Against Stress

By contrast, the type of behaviour denoted by segmentation can be seen in the Works Committee, many of whose members shut themselves off from the shop floor as a means of avoiding contact with the unfavourable attitudes common among operatives to those who seek elected positions. Works Council members expressed their annoyance at the dis-interest of their constituents by adopting the attitude that it was not their business to communicate with them—if their constituents wanted to know what was going on then they should take the trouble to sit in the Strangers' Gallery and find out for themselves. In the Superintendents Committee, stress, both within the Committee and between the Committee and higher management, created for some time a considerable barrier to effective communication which had a disruptive effect on executive work that is only recently disappearing as a result of working-through stresses in the executive system.

It is for such reasons that disruptive failure of communication is regarded not so much as a matter of inadequate organization, or of fortuitous circumstances, as an unconscious but highly motivated type of behaviour intended to obviate stress and discomfort but producing in effect social segmentation. As a defence technique, segmentation can only work up to a point; for either the segmentation is carried so far that the unity of the organization is impaired, or the feelings and attitudes which are not directly expressed seek indirect outlets. It is virtually impossible to avoid communicating feelings and attitudes to others. For instance, higher management was anxious over the reception of its social policy by its next line subordinates. Influenced by this, they erected a kind of controlling dam to prevent two-way discussion of such matters within the executive. Instead they talked over the ideas in question at Works Council with the Works Committee. There were many indications, however, that the meaning of this behaviour was not intended to be missed, nor was it missed in the lower ranks of management. Consciously, recognition may have been lacking, but there was present an intuitive feeling of resentment at being isolated and by-passed. Innumerable examples of the indirect expression of attitudes were to be found in recurring complaints about off-centre targets for hostility, such as the canteen, by means of which a host of pent-up feelings were getting an indirect outlet.

Conditions for Effective Communication

Effective communication requires a known and comprehensive communications structure. In Glacier this is represented by the executive, consultative, and appeals channels. Secondly, there is required a code governing the relations between people occupying various roles. This is represented by the newly developed Company Policy document and by the customs and procedures which while unrecorded are nevertheless a part of accepted practice. Finally, there is required a quality of relationship between people immediately connected with each other such that adaptive segregation may be mutually agreed, and stresses worked at so that rigid segmentation becomes unnecessary.

It is with regard to the last of the above three conditions that most trouble arises, as can be seen if the data on communications in the executive system are examined. This takes us back to problems related to the T-group structure of the nuclear commands. In the case of a communication from higher management to shop floor, a breakdown anywhere along the line reflects in some degree a breakdown within the top management group. There were many examples of instructions passed from the General Manager to his divisional managers for dissemination down through the executive lines which did not have their finally intended effects. In such cases, rather than scapegoat their subordinates, top management have learned to accept responsibility themselves. Either the General Manager had failed to pass on the full weight of his authority, so that the communication would have sufficient momentum to pass down, or there had been a failure in the upward flow of information—the divisional managers had not ascertained the causes of the problem, or the needs of the personnel lower down. For communications to operate effectively in an hierarchical channel, there must be freedom from anxiety at the top, and a willingness not only to receive upward communications, but essentially to exert a continuous pull to ensure that subordinates bring crucial problems forward. This they will do only when they have the knowledge that they will get help and support, and, where deserved, criticism of a fair and constructive kind.

GROUP TENSIONS AND WORKING-THROUGH

A social structure which fits its work task, and an effective system of communications, are basic requirements for adaptation and change. But strong forces were working against adaptation and change, in the form of unconsciously motivated segmentation and role confusion. It has been the goal of the Research Team to help groups in the factory to get at the nature of these forces as a means of enabling them to take better decisions concerning changes they needed to make. The general hypothesis was sustained that the factory as a whole would become more flexible through the experience of successfully tackling a small number of problems in a comprehensive way. In this section we shall inquire into the conditions and means of working towards change.

The Method of Working-Through

The research task set was that of developing methods of offering technical assistance to groups that requested the help of the Team in exploring underlying and concealed forces—whether psychological, cultural, structural, or technological—that were impeding their progress or otherwise reducing their efficiency. So far as this meant uncovering forces that had gone unrecognized either through being consciously or unconsciously ignored or denied, resistances were encountered. For to put up with the continuation of the problem was often felt to be less painful than to undergo the changes required for its resolution. The method used was to draw attention to the nature of the resistance on a basis of facts known to those concerned. Opportunities were taken to illuminate in the specific situation the meaning of the feelings (whether of fear, guilt, or suspicion) that constituted the unpalatable background to anxieties that were present about undergoing changes that were necessary. When successful, interpretations of this kind allowed group members to express feelings which they had been suppressing sometimes for years, and then to develop an altered attitude to the problem under consideration. Even awkward or over-blunt comments often came as a relief.

306

It was never possible to uncover all the operative factors, nor was this set as a goal. In the very process of studying a particular problem, new problems were uncovered or created. All that was aimed at was to obtain a sufficiently clear picture at any given time to allow practical advance to be made. There always remained undealt with a receding hinterland of unrecognized influences, old and new, emanating both from large-scale social processes that were affecting the affairs of the group, and from irrational and unconscious forces that were motivating the behaviour of individual members.

This process of helping a group to unearth and identify some of the less obvious influences affecting its behaviour is one which is borrowed from medical psychotherapy, from which is borrowed also the technical term *working-through*. It presupposes access by a consultant trained in group methods to a group accepting the task of examining its own behaviour as and while it occurs, and a group able to learn, with the aid of interpretive comment, to recognize an increasing number of the forces, both internal and external, that are influencing its behaviour. The expectation, then, is that the group will acquire a better capacity to tolerate initially painful insights into phenomena such as scapegoating, rivalry, dependency, jealousy, futility, and despair, and thence a greater ability to deal effectively with difficult reality problems. When we speak of a group working-through a problem we mean considerably more than is ordinarily meant by saying that a full discussion of a problem has taken place. We mean that a serious attempt has been made to voice the unrecognized difficulties, often socially taboo, which have been preventing it from going ahead with whatever task it may have had. There is entailed the development of an awareness by the group that tensions exist within it which are lowering its effectiveness; and implied is a willingness to undertake their open discussion as a means of accomplishing their resolution.

The resolution of stresses within a group is to be regarded as a continuous and unending task. It may not be noted as an item on the agenda of any committee, yet every item taken involves some work of this kind. Once a group has developed insight and skill in recognizing forces related to status, prestige, security, authority, suspicion, hostility, and memories of past events, these forces no longer colour subsequent discussion nor impede progress to the same extent as before. Dealing with them accounts for a smaller part of the group's activities, absorbs less of its energies, and allows it to handle more effectively those issues which are on the written agenda.

The Service Department: An Example of Working-Through

In the negotiations in the Service Department on the change-over in methods of payment, the early attempts of the management and of the workers' representatives to keep discussion narrowly focused solely on the wages question came to naught. Each time they tried so to confine themselves and to "stick to the agenda", other issues crept in, causing intense frustration to everyone and creating the feeling that they were never to get anywhere. Eventually it became apparent that these so-called side-issues were not so unrelated to the problem at hand as at first they seemed. They contained in fact problems of relationship which had to be tackled before the wages question could be successfully approached.

First and foremost were the unresolved relationships among the members of the negotiating group: the shop management, the supervisor's representatives and the shop committee members. A severe mutual test-out occurred. Management took every opportunity of allowing the workers to demonstrate how responsibly they were capable of behaving. The workers did not miss a single occasion to test whether the management really intended to be as co-operative and above-board as they professed. The fundamental issue was whether relations based on mutual confidence could be established, and as the testing-out of this issue progressed, so did the tempo of the discussions. The conclusion of one phase came with the acceptance of the proposed Shop Council as an institution through which many problems that had arisen could be resolved once the change-over had been implemented.

The function of the consultant was to make such observations as would help those concerned to recognize some of the less apparent factors affecting the immediate situation, particularly the stresses in the face-to-face group. There was resistance to his interpretations, varying from polite denial or ignoring of his remarks to expressed annoyance or irritation. Where he felt justified, he used these reactions to himself to show the group how uncomfortable they felt about their own relationships, and that they were transferring on to him, as the person bringing these problems into the open, their own attitudes towards each other. In this way the consultant was able to assist Service Department members to explore many of their involuntary attitudes: the workers' suspicions of the management, that arose partly out of personal reactions, partly out of departmental history, and partly out of more traditional attitudes of workers towards industrial management; anxieties of the Divisional Manager and his staff that the workers and their representatives were likely to behave destructively and irresponsibly; the fears in the department generally of being dominated by higher management and by the main works; the mistrust felt by the workers' representatives of each other, and the lack of real, as distinct from superficial, unity among the workers in the shop. These attitudes were not talked about in a general way, but were taken up on the spot, as and when they occurred, so that the negotiating group could share the direct and active experience of feeling what it was like to be testing each other out on these scores, when to all intents and purposes they were having a rational discussion of a wages problem.

It was possible for the consultant to help certain groups to experience their difficulties only because of a strong desire that was present in them to find a common ground on which to establish constructive relationships. This was a point which was frequently lost sight of, and it was necessary from time to time for the consultant to show a particular group that it was discovering how good its relations really were by testing the limits of how far they could be strained without breaking. Once it had begun to forgo such intense anxieties and suspicions through having tested their reality, a group became more able to accept the goodness of its own relationships. The possibility then emerged of rapidly coming to grips with the external realities of economics and departmental administration in settling a wages question such as that which arose in the Service Department. Discussion was protracted because those concerned chose to pursue the matter to the point where their relationships were sufficiently sound to allow them to take practical decisions with relative ease. It is this time-consuming and disturbing process of working-through which ordinarily causes so much impatience. But, taking the Service Department experience in its own right, it seems fair to conclude that initial impatience was a reflection of resistance to facing problems of group relations.

As some of this resistance was resolved, so too did impatience disappear.

Factors Allowing for Working-Through

Two interrelated factors are necessary in any successful process of working-through of group problems, and a third is desirable and frequently present. The necessaries are a group with a problem severe and painful enough for its members to wish to do something about it; but also of a sufficient cohesion of purpose, or morale, to render them capable of tackling it and of seeking and tolerating necessary changes. It is this combination of pain and morale that induces understanding acceptance of the illumination of difficulties unconsciously concealed because too devastating to admit; but it is only the giving of genuine understanding of such difficulties that permits resolution of the underlying stresses and their symptoms. The third factor which is a desirable and frequent accompaniment is the frustration experienced when there has not only been denial of the causes of difficulty, but when a stage has been reached where the denial mechanisms being employed no longer serve their purpose of giving relief from distress. A group in this state is ready for self-understanding, being unable to continue evading the issue.

The role of the Research Team in the working-through of group tensions stemmed from the independent relationship which it had established with the factory. This relationship was based on the principle of offering help only towards the goal of increasing the insight of those concerned into the reasons for the decisions they were making, or the difficulties they were encountering. Never was advice given, or judgment pronounced, as to the value of one course of action compared with another. To maintain this attitude was most difficult behaviour to learn, but it was noticeable that most assistance was given when Research Team members achieved detachment from any personal desire that groups in the factory should do this or that. The helpfulness of having present someone detached, in the sense of not being in favour of any of the possible outcomes of a contentious argument, was frequently commented on, even though his contributions may have been painful through bringing in the very issues being evaded or denied.

Working-Through and Adaptation

A pertinent question is how far working-through based on the direct examination in the face-to-face situation of the causes of stress in and between groups was necessary in order to produce the changes that took place. Would not straightforward advice on a new constitution for the Works Council or the Superintendents Committee, or a new organizational scheme for the executive system, have accomplished as much in shorter time? No definitive answer can be given to this question. Changes would have gone on in the factory whether or not there had been a research project. Many changes took place in regions outside the area of research activity. But where the changes occurred in collaboration with the Research Team, they were different both in manner and effects from what would have been the case without such collaboration. For example, recommendations could have been made to the Superintendents Committee for a number of different administrative solutions to the original problem of the committee chairmanship. But, as it turned out, chairmanship of the committee was only one small facet of the much larger problem of executive leadership in the management as a whole. The eventual solution came about only after a considerable change had taken place in the factory generally in the way managers (starting at the very top) perceived their functions, not intellectually but in their actual behaviour.

It is because it was the goal of the Research Team to assist members of the organization to bring about changes in their real behaviour, at a rate and in a direction that remained under their own control, that so much time was required and such difficulty met. Changes in real behaviour do not take place as quickly and as easily as do changes in professed attitude or in administrative arrangements. But these have less effect on what people actually do. People behave as they do for motives and purposes of the utmost importance to them. Though some of these are conscious, others are not; and if some go well together, others conflict. To assist in bringing about rational and planned changes in group relationships that would change real behaviour of those concerned, meant, first and last, assisting members of the firm to obtain a clearer understanding of their motives and purposes, and of where these were in conflict and where they were not.

To solve immediate practical problems for the factory was therefore inconsistent with the aim of the project, concerned as this was with developing methods of assisting groups to deal with such problems for themselves, so that the organization as a whole could become a more adaptive and self-dependent industrial community. It is doubtful whether giving advice to people on how to solve their problems can go far towards accomplishing such an end. It will be some time, however, before it can be determined how far the present work-through methods have gone towards creating that degree of insight which will allow a higher degree of conscious determination of policy to be regularly maintained. At least this is certain: that the present research has in no way proved a panacea for removing difficulties from the organization. The exploration of the interacting forces producing specific complaints, while often making rapidly possible the resolution of the complaint itself, also caused other hitherto unsuspected difficulties to be turned up. These were often of a deeper and wider character, with no prospect of rapid remedy.

It might be said that as things got better they also grew worse, and to those who suffer their own change it may seem at times too much like this. Yet when fundamental sources of stress are being dealt with, it may be discovered that solutions of multiple problems are brought within reach by the thorough working-through of a specific case. Once experience can indicate that a basis may be laid for the prevention of subsequent difficulties by the more thorough working-through of present discontents the additional pain and time demanded will be more easily accepted.

The fundamental problem confronting Glacier as a factory is not merely to ride the difficulties of the moment, but to maintain a social structure, culture, and corpus of personnel, which will allow it successfully to meet the changes demanded of it in a future in a changing society. Our definition of social health in an industrial community would picture a factory not so much free from problems, as one capable of tackling in a realistic way whatever technical, economic, and social problems it may encounter. To do this, the irrational and less obvious factors influencing its behaviour require periodic identification and working-through in order to protect it against piling up a slag heap of unresolved issues to bedevil the perception of the changing realities of the future.

SUMMARY: A GENERAL INTERPRETATION OF GLACIER'S DEVELOPMENT

In this final section we shall summarize the experience of the project in a general interpretation of the development of Glacier as a factory. The interpretation will consist of bringing toge*her the most significant events and factors which have interacted to give the present situation in the firm its particular character. A balanced assessment of the work so far done is not yet possible, for the project is still continuing. Evaluative studies are being carried out and will be reported subsequently.

The Continuing Effects of History

In order to understand the present situation, it is necessary to consider the dynamics of certain outstanding historical events in the life of the organization, events whose definite mark is left on the pattern of the present. There was at first a small metallurgical works in which the alloying of white metal was supreme, and engineering activities relatively unimportant. After 1920 the firm expanded very rapidly. 1935 saw an engineering outgrowth, with nearly 500 employees, superimposed upon the tiny white metal alloying and selling concern. But white metal remained master because of the interests of the owner who never became enthusiastic about the engineering developments. As a result, technological growth far outstripped efficient organization, since organizational methods had been left to develop in a haphazard way.

By 1935, although the white metal activities had continued unchanged, the new engineering development was stretching to bursting point its boundaries as minor partner in the concern. Radical changes occurred between 1935 and 1940. Engineering became dominant. It swamped the white metal work (which now remains as a small section of one manufacturing shop) and forced sweeping organizational changes. The firm was made into a public company, with the previous owner as Managing Director. He retired in 1938, and was replaced by co-managing directors. One of these was the Works Manager, a "production man" of long experience, and the other the Sales Manager, a younger man interested both in new technical and administrative methods.

The legal requirements for costing, stock and other financial controls through becoming a public company brought an extension of specialist controls. This trend was further stimulated by the relationship established with an American concern whose managing director was especially interested in advanced costing methods. Accounting, production control, production engineering, and personnel departments were introduced. A special research section was created. And beginnings were made at establishing a formal managerial structure, and stepping up managerial skills by means of training. All this happened during the period when there occurred an open split in the management. This split was precipitated by a difference in technical and social outlook between the co-managing directors. Although the older of the two resigned in 1939, the split remained. The new specialist functions were in the hands of those who were identified as supporting the younger man who became sole managing director, and who favoured the wider use of specialists. The line managers feared they might be dominated by the others. This fear was reinforced by the fact that greater controlling power was given to the specialists as a means of coping with the large numbers of new and inexperienced supervisors created during the wartime expansion. At the same time, the national emphasis on joint consultation in the engineering industry, combined with the personal interests of the Managing Director, led to the setting up of joint consultation in the firm. This laid the basis for an extension of democratic control. But it also added to the confusion about the functions of the executive, a confusion which had already been exacerbated by the split between the line managers and the specialists.

Through the general activity and growth due to war production, the 1940's saw the firm flourish and prosper technically and economically. These were also the years during which were resolved the mass of organizational problems left over from the rapid and unplanned technological development and physical expansion from 1920 to 1935.

Functional—Executive Split

The situation arising out of the split between the so-called functional managers and executive managers was further complicated by the growth of the concern into a multiple organization. It was not seen that, as a result of opening new factories, the Managing Director had taken on an entirely new role in addition to that of executive leader of the London factory. This threw a greater load on to his London subordinates, a load that was difficult to handle because the factory did not have a clearly structured executive system. That these difficulties occurred and persisted was partly the effect of the motivation of the top executives. They maintained a grip on their subordinates by making one responsible for checking-up on another—a division of responsibility and authority that caused confusion and some resentment. But there was also an unconscious co-operation—and even collusion—between all members of the management system, to maintain just the right balance between confusion and efficiency in executive relationships; enough efficiency to keep the firm in a satisfactory economic position; but enough confusion to avoid too clearly identifying the full executive responsibilities and authority of individuals.

As against the difficulties thus depicted, there was also an intensive working out of relationships among managers. This was consequent upon getting out a reasonable organizational chart for the factory—a job which took three years to complete and which laid the foundation for subsequent profound changes. But the completion of an organizational chart did not by itself automatically cause managers to take up the leadership and control of their subordinates. The culture of the organization had been fashioned to act as a defensive system against the anxiety of having to have too explicit a responsibility for people. It was possible, for instance, for superintendents, over a period of years, to meet on their own without at the same time either having the opportunity, or being required, to meet regularly as a group with their own executive chiefs.

As these problems were thrown into relief in such groups as the Works Council, the Divisional Managers Meeting and the Superintendents Committee, changes in the executive structure and practices took place, and are still in progress. It was possible for these changes to take place more easily because of the existence of the factory organizational chart, and because of the knowledge of managerial principles possessed by a number of managers. But neither of these factors was of much help in overcoming the emotional attitudes which were hampering executive work. The Managing Director had first to accept his additional role of General Manager. This meant emotional acceptance of the fact that he was the executive leader of the divisional managers. Once accomplished, this change gave the factory continuity in the executive lines at the top. This process was accompanied by a working-through of some aspects of relationships within top and middle management, so that the executive lines at mid-management level were put straight. A policy document could then be got out, in which these changes were recorded (and this policy is being actively elaborated and, evolved). The emphasis is now on providing a coherent executive system extending right through to the shop floor, and based on sanctions which include democratic discussion through consultation.

Executive-Consultative Confusion

The use by higher management of so-called functional managers, to keep control over the line managers, was reinforced by the use of the joint consultative machinery to by-pass the executive chains and establish direct contact with the workers' representatives. These latter were regarded as a more reliable source of information about shop floor matters than the middle and lower grades of management. The Works Committee and Works Council were set up without the support of the middle management levels. Nevertheless, the meetings between workers and top management made it possible to put into practice, to some extent at least, the principle that people should be able to participate in the decisions that affect them. The existence and growth of the consultative system marked a change in the character of the firm. It established a basis for good relations between workers and management, and for tackling some of the cynicism and despair left over from the confusion of the 1930's. With the development of the *Principles of Organization*, hammered out by the Works Council, the pattern was set of the Council taking part in policy making, while leaving to the management the implementation of policy. At the same time, however, joint consultation had the effect of increasing the sense of loss of authority experienced by the line executives. The shop floor had a channel round them. Instead of supplying executive leadership for the subordinate management grades, top management was getting together with the workers' representatives. The fact that the superintendents, through their committee, had from the beginning been given a seat on the Works Council did not improve this situation. On the contrary, it served to exaggerate the difficulties by giving middle management the feeling that their contact with top management was no closer than that of the workers (if indeed as close).

The difficulties thrown up by joint consultation arose from unconscious motives which were at play along with the conscious and constructive aspects of the developments. The top management were partly evading their anxiety about facing and work-·ing-through their relationships with each other and with their subordinates. And the Works Committee representatives were using their relations with higher management, whether co-operative or antagonistic, to avoid having to face their constituents. Both these groups also mirrored larger scale forces in the factory. Everyone desired to avoid stresses in group relationships; and the Works Council provided the lead that was wanted—a lead in flight from dealing with these stresses.

This consultative collusion between the appointed management, and the elected staff and workers' representatives, was bolstered up by a number of strongly held beliefs. One of these was the general belief in industry that if only the discord between management and workers could be resolved, everything else would come out all right. Another was the belief that irrational or emotional influences could not be at work among people who had reached the level of top management or had become elected representatives. But the beliefs that caused the most argument were those based on ultimate values associated with principles of democratic living. For instance, the view that if people were treated responsibly, they would behave responsibly, was adopted (although individuals differed in the whole-heartedness of their endorsement of such a view). That these beliefs contained much truth, there is no doubt. But the element of truth was challenged by critics, whenever the beliefs were put forward—as they often were—as the whole truth.

The need to erect such a structure of beliefs was partly related to uneasiness over accepting—in one's consultative role—the fact that people could be driven by destructive as well as constructive, materialistic as well as idealistic, impulses. To obviate the need to face the destructive impulses, both in themselves and in others, the existence of such motives was in part denied. This partial denial made it possible for those who were opposed to joint consultation to go to the other extreme, and to deride the attempts to extend democratic control as too philosophical and not sufficiently related to the "real" problems of producing bearings and earning a living. By working-through these less obvious causes of trouble, the Works Council members have been enabled to get a firmer, and at the same time more flexible, grasp of their practical goals for joint consultation. In particular, they have been able to conclude their clarification of the function of the Council as a branch of the policy-making network (a process of clarification which began in 1945 with the Principles of Organization), and to modify the structure of the Council to one potentially better suited to the new demands. In so doing, the relationship between consultative and executive functions was straightened out, leaving sharply silhouetted the then central problem of the operation of the executive system.

Aspects of the Present Position; December 1950

The constructive attitudes of those who brought about changes in the factory were sufficiently tough and resilient to survive the difficulties encountered. The developments in the executive structure and joint consultation had combined to lay the basis for solid morale in the firm. It was this morale which made it possible for the members to tackle problems in a comprehensive manner, as well as at great depth. In this way, the consultative-executive collusion, and the collusions within the management system, both of which were the results of the particular way in which the organizational confusion of the 1930's was dealt with, were worked at and partially resolved between 1947 and the present time.

On the executive side, what is now the General Managers Meeting is deeply engrossed in working out a policy governing executive responsibilities, including the role of the manager, and the conditions governing executive and functional relationships. The first draft of the new policy will be found in *Appendix II*. The General Managers Meeting have drawn up this policy in terms of themselves, and their problems as a group. Hence they have been able to advance a policy for the executive system, consistent with their own needs. This policy is gradually being worked-through in the executive system as a whole. The Works Managers Meeting are tackling the problem of training for them-selves and for their subordinates the foremen and section super-visors. A working conference is scheduled for the Works Division executive chain, from the General Manager down to the Superintendents, to discuss the executive policy, in particular the authority and responsibilities of the executive, as a prelude to the introduction of supervisory training. At the same time the

Works Manager has been meeting each superintendent and his subordinates, to discuss departmental executive policy and organization. There is also growing up, in the factory as a whole, the practice of each manager meeting regularly with his own nuclear command. Consideration is being given to overcoming the practical difficulties of extending this practice to shop floor level.

On the consultative side, the new Works Council has had its first meetings and is engaged in settling down. It is too early, as yet, to ascertain the pattern of Council work which is likely to emerge. One striking event may, however, be mentioned. The General Manager is now presenting a monthly management report to the Council in place of the multiple reports from the divisional managers. He is able, by this means, to give a monthly integrative report on the state of the factory, in which he outlines all the major management actions. This report has an impelling and authoritative quality because of its overall coherence, a coherence which was impossible to obtain when the divisional managers reported separately. The new Works Committee of shop stewards, which came into existence just one week before the Works Council, is also settling down. The Works Committee, and the three Staff Committees, are all now faced with the immediate task of building an effective relationship with the Works Council.

In short, at the turn of the year, 1950-1, the factory is engrossed in completing one phase of the task of creating an executive system extending continuously from Managing Director to shop floor. The authority of the executive system will be based on a known policy forged within the factory and therefore indigenous to it. Helping to sanction the evolving executive procedures is the new consultative set-up in which policy governing executive action is hammered out. This consultative structure is now clearly distinguished from the executive structure. The General Manager's monthly report assumes considerable importance as providing the main link between the factory-wide executive and consultative systems. The management, by bringing a regular report on its work, leaves itself open to criticism and to challenge on policy concerning all aspects of factory activities. In this way the management receives constantly renewed sanction for its work from all sections.

At the same time the firm continues its physical expansion and technical growth. Two large factories are under construction. A recent new share issue was substantially over-subscribed. The big remaining question is whether history is to repeat itself in a phase of technological development, proceeding independently and leaving organization stresses to be caught up with later; or whether the experience of the past years has struck home sufficiently deeply to allow the social-technical split to be eliminated, and for organizational flexibility and skill to accompany and facilitate technological health and development.

APPENDIX I

COMPANY POLICY

This appendix reproduces the Company Policy document which was drawn up by the Managing Director in collaboration with the senior management groups of the two factories, and issued during October 1950. The document has been accepted by the two Works Councils as a working policy pending any modifications which may arise as a result of current discussions within the executive and consultative systems in both factories.

SECTION A: DEFINITIONS

A.1. *COMPANY POLICY is that Policy which affects the activities and members of the Company's factories. It shall derive from the Board of Directors. It is divisible as follows:*

A.1.1. *DEFINITIVE POLICY is that part of Company Policy which is vested in the Board of Directors by the requirements of the Memorandum and Articles of Association of the Company, Company Law or other legislation, and the regulations of the Stock Exchange.*

A.1.2. *CONDITIONAL POLICY is all policy which can be made the subject of consultation with Factory Councils without prejudice to the requirements, statutory or otherwise, imposed on the Board of Directors.*

A.2. *The Policy governing the ADMINISTRATION OF THE COMPANY will therefore be in part Definitive and in part Conditional, and shall be built up in writing from:*

A.2.1. *Policy decisions already existing,*

A.2.2. *Policy decisions arising naturally out of high-level appeal decisions made in the past, accepted by General Managers, that have not been challenged.*

A.2.3. *Policy decisions which arise by historical precedent, until such time as these be questioned, and/or their place taken by explicit decisions.*

A.3. *Any member of the Company shall have the right to initiate AMEND-MENTS TO EXISTING POLICY, provided such proposals are made to the Board through the duly constituted channels.*

A.4. *DEFINITIVE POLICY, being largely confidential to the Board of Directors, will be constituted and recorded in the Minutes of the Board. It will be communicated by the Managing Director through normal executive and/or representative channels when permissible and in the Company's interest.*

A.5. *CONDITIONAL POLICY shall be decided by agreement between the Board, normally represented by the Managing Director, and the appropriate representative bodies of the Company's factories, e.g. factory Councils or factory Staff Councils.*

Appendix I

SECTION B: THE PURPOSE OF THE COMPANY

B.1. *THE PURPOSE OF THE COMPANY, and those working in it, shall be the continuity and expansion of a working community, the conditions of which will enable its members to serve society, to serve their dependants, to serve each other, and to achieve a sense of creative satisfaction.*

B.2. *This purpose will best be ACCOMPLISHED BY:*

B.2.1. *Seeking the maximum technical efficiency;*

B.2.2. *Seeking the utmost possible organizational efficiency;*

B.2.3. *Seeking to establish an increasing democratic government of the Company community, which will award fair responsibilities, rights and opportunities to all its members, consumers, and shareholders;*

B.2.4. *Seeking at all times to earn such revenue that the Company will be able:*
To provide such reasonable dividends for its shareholders as represent a fair return on their capital investment for the speculative risk incurred. To undertake research and development in order to enable the Company to attain a high position in the competitive market.
To provide those who work in the Company with working conditions which will promote their physical and mental well-being.
To improve its equipment to enable those who work in the Company to do so with the greatest possible effectiveness.
To raise wages and salaries in order that those who work in the Company will be able to live full and happy lives.
To improve the Company's service to its customers by reducing the price or improving the quality of its products.
To make reserves to safeguard the Company and those who work in it.

SECTION C: FINANCIAL POLICY

C.1. *Provided the shareholders receive a reasonable return on their investment, the SURPLUS REVENUE of the organization shall be used only as laid down in Section B.2.4.*

C.2. *The PRIORITY given to these objectives shall be governed by the circumstances which obtain from time to time.*

C.3. *The efficiency of units of the Company shall be judged by the following criteria:*

C.3.1. *The profit made.*

C.3.2. *The out-put per man-hour compared with that achieved by other units of the Company performing similar tasks.*

C.3.3. *The out-put achieved compared with the theoretical potential.*

C.3.4. *The labour turnover compared with that of other similar organizations.*

C.3.5. *The results shown by the financial and statistical statements referred to in Clause C.5.*

C.3.6. *Other measures of efficiency agreed from time to time.*

C.4. *The financial result of CHANGES IN METHOD which increase efficiency or output shall benefit the Company as a whole, and not only those who devise or operate the improved method.*

C.5. *FINANCIAL and statistical STATEMENTS comparing results achieved with those budgeted shall be available to all executives whose sphere of responsibility is large enough to warrant its treatment as a cost centre.*

322

C.6. *All proposed expenditure shall be subject to controls which will be specified from time to time by the Directors or otherwise under the authority of the Board.*

SECTION D: FACTORY COUNCILS

D.1. *Each factory shall, and each major department or unified group may, establish BODIES REPRESENTATIVE of Managers and Works and/or Staff members. The Senior Manager present shall carry the responsibility of representing the higher policies within which such bodies must operate.*

D.2. *CHANGES in or additions to POLICY shall only be valid if agreed unanimously by these bodies in formal meeting. These bodies shall be called Factory Councils, Divisional Councils, Departmental Councils, Staff Councils, etc., as is appropriate.*

D.3. *Factories and Departments of the Company shall be MANAGED BY COMPANY EXECUTIVES within the terms of Company Policy, and within the terms of Factory and Departmental Policy.*

D.4. *No joint policy-making body shall agree any change or addition to policy which may affect people working in any part of the Company outside the SPHERE OF JURISDICTION of the body concerned.*

D.5. *Any member of a policy-making body may CHALLENGE a DECISION taken by that body, or by a joint policy-making body at a lower level in the joint consultative structures. A decision so challenged shall be referred to the joint policy-making body at the next higher level in the joint consultative structure.*

D.6. *The following GENERAL PRINCIPLES shall govern the set-up of bodies representing members:*

D.6.1. *Each primary working group shall elect one of their members to serve with a secondary group of representatives. If the size of the unit warrants it, the secondary group shall elect to a tertiary group of representatives. By this means, an employee representative structure shall be built up which is collateral with each major level of executive authority.*

D.6.2. *Formal meetings between managers and these representative bodies shall take place for the purpose, among others, of amending or adding to policy. Due notice of such meetings shall be given to all members of the organization or part thereof concerned, who shall have the right to attend as audience whenever reasonably possible.*

D.6.3. *Managers shall do everything possible to keep these bodies informed of the current situation and future plans of the organization. No facts shall be denied to them, provided that the interests of the organization or its members will not be damaged by their publication. It shall be the duty of each representative body to keep all their constituents fully informed on all transactions at meetings.*

D.6.4. *These representative bodies shall be given all possible means and safeguards to enable them freely to present their criticisms, suggestions and aspirations and those of the people they represent.*

D.6.5. *In the event of failure of a joint policy-making body to agree unanimously a proposal of policy, it shall be the right of any member or members of that body to require that the matter shall be referred to the next higher level in joint consultative structure.*

Appendix I

D.7. The responsibility and authority for making all necessary DECISIONS FOR THE IMPLEMENTATION of Policy shall be vested in the executive management of the Company.

D.8. Every executive shall, whenever possible, give the REASONS FOR DECISIONS and instructions issued by him.

D.9. Any member of a joint policy-making body shall have the right, at formal meeting of that body, to QUESTION the EXECUTIVE ACTION of a manager. An action so questioned shall be the subject of formal discussion. In the absence of any such challenge, executive action shall be regarded as agreed by the representative body concerned.

D.10. In addition to the formal consultation outlined above, there shall be, at all times and between all levels of the executive and representative structure, the fullest possible INFORMAL CONSULTATION.

D.11. If the highest policy-making body in any one of the Company's factories is unable to reach agreement on a major principle, then the matter shall be REFERRED TO A JOINT MEETING of representatives of the Company's Factory Councils and of the Board. Failing agreement at such a joint meeting, the matter shall be referred to the Board of Directors.

D.12. It shall be the RIGHT OF INDIVIDUAL MEMBERS of the Company to decide whether or not they shall join an appropriate Trade Union.

D.13. Although believing that it is in the best interests of the Company that members should join their appropriate Unions, it is deemed improper for Management to use any INFLUENCE TO PERSUADE members of the Company to seek Union membership.

D.14. The right to determine change in the CONDITIONS FOR THE ELECTION OF REPRESENTATIVES shall be vested in members of the Company.

D.14.1. Such change is only valid if submitted to BALLOT by those who will be directly affected, and supported by not less than two-thirds of the valid votes cast.

D.15. Management shall carry a general responsibility for refusing to accept the result of a ballot, should there be adequate evidence that IMPROPER PRACTICES have or may have been used. This responsibility shall be exercised through their membership of Works Councils, and other bodies, and by virtue of their share in the responsibility for the appointment of Returning Officers.

D.16. Because the structure of representative committees is of vital importance to the efficient running of the Company, any proposal for the amendment of conditions of election or Constitution of representative bodies shall, wherever possible, be the SUBJECT OF COLLABORATION between Management and those seeking the amendment.

D.17. One representative nominated by each of the Works Committee and Staff Committees at the Company's Factory Groups, both in London and Scotland, shall hold one Ordinary Share of the Company, in his own name and in his own right, thereby entitling him to REPRESENTATION AT ALL MEETINGS OF SHAREHOLDERS of the Company, in accordance with the terms laid down.

D.18. Such representatives, by virtue of their shareholding, may attend meetings of shareholders, so that the views of the Committee they represent concerning *the agenda before the shareholders' meeting may be expressed in the normal way open to shareholders.*

SECTION E: INDIVIDUAL RIGHT OF APPEAL

E.1.   *Every member of the Company shall have the RIGHT OF APPEAL to successive stages of higher executive authority. Every effort shall be made by all members:*

E.1.1.   *To make known to individuals their right to these channels of appeal;*

E.1.2.   *To prevent victimization of those who use them;*

E.1.3.   *To deal with appeals as speedily as possible.*

E.2.   *At the departmental stage, an Appeal may be referred by either party to the Factory Personnel Department, for IMPARTIAL RECOMMENDATIONS to the appropriate executive authority and/or the appellant. Should his recommendations not be satisfactory to both parties, the appropriate Personnel Officer may refer the appeal to any level in the executive hierarchy, but shall notify, and, if desired by the executives of a particular factory, seek the agreement of all INTERMEDIATE executives, who shall be entitled to ATTEND the hearing and submit their views.*

E.3.   *Appeals against the decision of the Managing Director which cannot suitably be referred to National Arbitration procedures can, with the approval of a body set up by the appropriate Factory Council, be referred to a COMPANY APPEAL TRIBUNAL. This Tribunal shall consist of one representative of Management, one representative elected by the employees, and an independent Chairman appointed on the recommendation of an appropriate legal professional body. The majority decision of this Tribunal shall be final and binding within the Company.*

E.4.   *At the HEARING of ANY APPEAL:*

E.4.1.   *The appellant, his chosen representative, if any, and the person against whose decision the appeal is made, shall all have the right to be present. Where it is in the interests of the Company that an Appeal be heard, and the appellant fails to attend or to appoint a representative, a representative shall be appointed by the officers of the Works Council, and the appeal shall proceed.*

E.4.2.   *The appellant shall have the right to be represented by any member of the Factory he may choose.*

E.4.3.   *The executive concerned shall encourage the appellant to have present an officially elected representative.*

E.4.4.   *Both parties can, with the permission of the person hearing the Appeal, call witnesses in order to establish facts relating to the circumstances.*

E.4.5.   *The decision shall be given in the presence of both parties.*

E.5.   *JUDGMENTS given BY THE MANAGING DIRECTOR on matters relating to the interpretation of Company Policy shall, unless challenged, have the standing of precedents which can be cited or used for the purpose of future appeals.*

E.6.   *When the resulting INTERPRETATION OF POLICY is not acceptable to individual members of the Company, then such members must petition either the body through which they are represented, or a member of a Work*

Council, to table the amendment before that Council. Such amendments, if agreed by the appropriate bodies, shall not affect previous judgments given.

E.7. Decisions by the Managing Director which, in his opinion, involve IMPORTANT INTERPRETATIONS OF POLICY, shall be COMMUNICATED to the Factory through executive channels, and a file of such interpretations shall be set up for the use of those hearing appeals in the future, and shall be open to all members of the Company.

E.8. Decisions given by Divisional or Senior Managers which are not challenged and which appear to them to be important interpretations of policy, shall be communicated to the Management Group of which they form a part.

SECTION F: MANAGEMENT STRUCTURE

F.1. There shall be one CHIEF EXECUTIVE (Managing Director) of the Company, who shall be responsible for carrying out Company Policy. He shall be a full-time worker in the Company, and shall have authority to take any action necessary for the implementation of Company Policy.

F.2. The Chief Executive shall be the only person having direct executive authority in all factories and branches of the Company. Members of the Board, (unless employed as executives, when they shall have the authority appropriate to the position), shall have no direct executive authority.

F.3. All possible means shall be used to explain to members of the Company the RESPONSIBILITIES of its EXECUTIVES, in general terms.

F.4. The SPAN OF CONTROL of a line executive shall be limited to the number of people he can effectively control, and amongst whom he can maintain co-operation. He shall be accessible to those immediately responsible to him.

F.5. The FUNCTIONAL AUTHORITY of all those carrying specialist responsibility shall be clearly set out and available to all members of the Company.

F.6. No member shall be EXECUTIVELY RESPONSIBLE to more than one person.

F.7. No member shall GIVE EXECUTIVE INSTRUCTIONS to any one other than his immediate subordinates, except in emergency.

F.8. DIAGRAMS clearly depicting the chain of command shall be exhibited in the factories.

SECTION G: THE MANNING OF COMPANY ORGANIZATION

G.1. The APPOINTMENT OF DIRECTORS is vested in the Shareholders.

G.2. The APPOINTMENT OF THE MANAGING DIRECTOR and SECRETARY is a definite responsibility of the Board of Directors.

G.3. The APPOINTMENT OF GENERAL MANAGERS shall be made by the Managing Director, in collaboration with a Selection Board which he shall appoint.

G.4. The APPOINTMENT OF DIVISIONAL MANAGERS and SENIOR MANAGERS shall be made by the General Manager, in collaboration with a Selection Board which he shall appoint. The Managing Director shall have the right to be a member of such a Board.

G.5. Executive, technical and supervisory APPOINTMENTS shall be made by methods acceptable to Factory Councils. Such methods shall take into account the following provisions of Company Operational Policy:

G.5.1. The opinion of the executive immediately responsible shall be regarded as of major importance when considering the suitability of an individual to fill a particular post.

G.5.2. Because the qualifications of executives materially influence the happiness of people, the selection of such executives shall be made by the most scientific methods available, and shall have due regard for the probable reactions both of his equals and his subordinates.

G.5.3. Provided suitable candidates are available, vacancies within the Company shall be filled by existing Company members. Such vacancies shall be advertised within the factory concerned, and in other factories of the Company at the discretion of the respective Personnel Manager.

G.5.4. The excellence of a man's performance in his existing job, or the absence of a suitable replacement for him, shall not be a valid reason for refusing him promotion to a post for which he is suitable.

G.5.5. Recognizing that frustration is caused by placing people in jobs which do not adequately occupy their faculties selection procedures will endeavour to find people whose mental and physical calibre is properly suited to the work involved.

G.5.6. Any member of the Company, through his elected representatives, shall have the right to make recommendations regarding the promotion of people within the organization.

G.6. Decisions concerning the TRANSFER OF MEMBERS OF THE COMPANY from hourly-rated basis to weekly-rated basis, or from one grade of staff to another, shall be made by methods acceptable to Factory Councils and/or to Staff Committees.

G.7. An executive shall have the RIGHT, subject to appeal, TO DISMISS a man from his own team. He shall not dismiss a man from the Company before conferring with the Personnel Department, in order to ascertain the possibility of providing alternative occupation. In the case of gross misconduct, executives at the level of Departmental Managers or above may dismiss a member, subject to appeal, without delay or prior reference to the Personnel Department.

SECTION H: CONDITIONS OF SERVICE OF MEMBERS OF THE COMPANY

H.1. The conditions of service of members of the Company shall be GOVERNED BY the following variable factors:

Normal Working hours
Rates of pay
Overtime allowances
Night shift allowances
Summer Bonus, if any, distributed as set out in Standing Orders, at the discretion of the Board of Directors
Travelling facilities
Holidays
Leave of absence

Appendix I

Payment during sickness
Pensions.

H.2.   Normal conditions of service for hourly-rated members of the Company shall be:

H.2.1.   Not less than the minimum standards determined by National Agreement, or by agreement between Trade Unions and the Federation of Employers.

H.2.2.   Holidays—as laid down in Standing Orders.

H.2.3.   Leave of Absence—as laid down by the provisions of the Company Sick Pay Schemes and in Standing Orders.

H.2.4.   Pensions—as laid down by the provisions of the Company's Pension Scheme.

H.3.   Variations in the rates of pay of hourly-rated members of the Company working on similar jobs shall be determined by ability or merit awards.

H.4.   Conditions of service for other members of the Company will be determined by the position held, responsibilities, and type of job performed. Taking all factors into consideration, a reasonable balance shall be maintained between all members of the Company. Where the conditions of such members are the subject of Union/Federation agreement, such agreed conditions shall be regarded as forming the MINIMUM STANDARD.

H.5.   In determining the conditions of service of any individual, the following FACTORS shall be TAKEN INTO CONSIDERATION:

H.5.1.   The value or potential value to the Company of QUALIFICATIONS and EXPERIENCE held.

H.5.2.   The degree of RESPONSIBILITY demanded by the job.

H.5.3.   SKILL in performance of the job.

H.5.4.   The possession of a PERSONALITY calculated to be helpful to the Company and its members.

H.5.5.   Willing readiness to render SERVICE to its members through any of the Company's institutions.

H.5.6.   PAST SERVICE of notable value to the Company.

H.5.7.   SPECIAL QUALIFICATIONS or attributes which enable the individual to command a high valuation of his services in the community.

H.6.   It shall not be considered improper for members of the Company on PIECE-WORK to earn more than staff or hourly-rated members, provided that:

H.6.1.   Piece-work prices are fairly rated.

H.6.2.   Such earnings are warranted by the skill and dexterity of the operator.

H.6.3.   The remuneration and other conditions of service of staff or hourly-rated members are just and adequate.

H.7.   Every member shall have the RIGHT TO EXAMINE the calculations which form the basis of his output standards; he shall be assisted by his representative if he so wishes.

H.8.   The conditions of service of all members of the Company shall be REVIEWED from time to time, to ensure as far as possible that a reasonable balance is maintained between individuals and groups.

H.9.   It shall be proper for any individual or group in the Company to bring to the notice of Management, through the proper executive or representative channels, cases where ADJUSTMENT OF TERMS OF SERVICE appears necessary.

H.10.   Any adjustment in the terms of service of groups within the Company shall be *introduced only after FULL DISCUSSION between all those who may be affected.*

H.11. *Members completing 21 YEARS OF CONTINUOUS or broken SERVICE with the Company shall be entitled to one week's extra holiday with pay to mark the occasion.*

H.12. *HOLIDAYS AND PAY ARRANGEMENTS will be at least equal to the minimum agreed between the appropriate employers' organizations and Trade Unions, and every effort shall be made to maintain a fair balance as between various units of the Company in the light of prevailing conditions.*

H.13. *Operating DETAILS and arrangements for holidays are the responsibility of the appropriate Factory Councils, Staff Committees and Departmental Councils.*

H.14. *Where the appropriate Factory body desires to change the time of a holiday from the normally accepted day to some different day, this shall only be done if the appropriate Trade Unions agree that the new holiday decided upon also becomes the day on which PREMIUM RATES OF PAY shall apply in the event of people working on that day.*

H.15. *Holidays of staff members or hourly-rated members shall be decided by the Factory Councils or Staff Councils in such a way as to safeguard Company efficiency.*

APPENDIX II

FACTORY STANDING ORDER ON POLICY GOVERNING EXECUTIVE BEHAVIOUR

The following is the policy document which was the first result of the working-through process which occurred during the special management development meetings which took place each week beginning in September 1950. Noteworthy is the adoption of the principle that executive responsibilities apply to everyone in the organization, and the resolution of the "functional manager—executive manager" split by the recognition that so-called functional management is not a type of management but a type of relationship between people in their executive roles.

DEFINITIONS

1. This order details many aspects of the formal behaviour required of people. It is not possible to designate which aspects are clearly new and which old, because no statement of previous required behaviour exists with which to compare the new.

2. In one particular respect, however, there is a clear change. In the past, the two classifications Executive Manager and Functional Manager have been used on the understanding that there are some who spend the majority of their time discharging executive responsibilities, and others the majority of their time rendering help by way of prescriptions and services. Examination of the jobs performed by people in functional divisions and departments of the Company fails to reveal any which entail a majority of time spent in giving prescriptions or services to others; there are some who possibly spend as much as 25 per cent. of their time in such activities, but they are very few in number.

3. It is extremely important, for the understanding of this Order and for the sake of better understanding in the factory, that we define clearly various types of relationships, roles and conduct. The remainder of this section is devoted to such definitions.

4. *The Executive System* is that part of the Company organization by means of which authority and responsibility are delegated and day-to-day work carried out. It is made up of a large number of positions, which will be called Executive Roles.

5. *Executive Roles.* Examples are Managing Director, Divisional Manager, Chief Metallurgist, Works Accountant, Superintendent, Foreman, Section Supervisor, Operative, Clerk, Inspector, etc. It is possible for one person to occupy more than one role, or for one role to be split up among a number

330

of people. When such dual roles or split roles are not recognized and clearly defined, this often gives rise to difficulties.

6. Note that the executive system extends down to and includes operatives, clerks, etc., a person being in his executive role while he is carrying out his excutive or job responsibility.

7. *The Structure of the Executive System* is depicted in the authority charts of the factory. This structure determines the relationships between various positions, thus the role of section supervisor is positioned at the head of a section of operatives and subordinate to an assistant foreman. The position in which a person is placed will force him to make formal relationships with certain other people who occupy roles which are related to his own. These formal relationships are called role relationships.

8. *Role Relationships.* There are two main types of role relationships in the executive system: line relationships and functional relationships. Line relationships are the vertical relationships between executive roles in the same line of command. Functional role relationships are relations between any other roles in the executive system.

9. In other words, executive relationships are those between a person and his superior and his subordinates. Functional relationships are those between a person seeking a service or prescription and another person who provides the service or prescription; the two people concerned shall be called respectively the *responsible person* and the *servicing or prescribing person.*

10. *A Manager* is a person who occupies that special type of executive role which carries responsibility for subordinates.

11. It is extremely important for the understanding of this Order to distinguish between formal and informal conduct.

12. *Informal Conduct* is governed by the unwritten codes of conduct that are part of the inherent make-up of each individual. When two people's codes are not identical, emotional differences may arise, and it is laid down in this Order as the formal duty of the executive to attempt to resolve such differences, but this Order does not deal with informal conduct. Indeed, any attempt to commit to paper the informal conduct required of managers would immediately cause it to cease to be informal. It is open to any member of the Company to make such informal contact with other members as he may desire, or as may assist him to carry out his duties, so long as he is clear, and makes it clear to others, that he is acting informally.

13. *Formal Conduct* alone is dealt with in this Order, and is the behaviour required by people in their line or functional relationships.

14. *The Nuclear Command* of a particular manager is that group of people for whom he is *directly* responsible.

15. *The Extended Command* of a particular manager is that series of nuclear command groups and persons for whom he is *directly* or *indirectly* responsible.

16. *A Prescription* is a technical recipe or direction which indicates what the prescriber deems to be the best method of achieving a desired result. Prescriptions are given by individual members of the Company with special knowledge in a particular field who have been charged by their executive superiors with the task of prescribing and of keeping abreast of new findings in that field of knowledge. Prescriptions can also be given by consultants who are not members of the Company.

17. Notwithstanding any responsibilities laid down in this Order, all Managers will be responsible for exercising normal common-sense in preventing what, in his judgment, would be seriously damaging or illegal acts, and in refusing to carry out instructions contrary to his own moral code, or to the Law or other overriding statutory authority.

EXECUTIVE RESPONSIBILITIES

A person in an executive role
18. Shall be responsible to his executive superior alone,
And shall carry responsibility
19. For the results of all his activities and those of any subordinates.
20. For reporting to his superior on his own actions and those of his subordinates alone, and not on the actions of any others.
21. For carrying out instructions received from his immediate superior, and for deciding what part of his responsibilities he delegates to his subordinates.
22. For informing the members of his nuclear command to what extent they are authorized to ask for services and prescriptions from others.
23. For knowing Company Policy and Standing Orders, and endeavouring to ensure that they are known and observed by his nuclear command.
24. For planning to meet the demands of present and foreseeable future situations.
25. For proposing such new policy or Standing Orders as seem necessary.
26. For seeking from the appropriate person those services which he requires to carry out his job, and which seem to him to be consistent with the Law, Company Policy, and Standing Orders.
27. For deciding, when he has been unable to obtain adequate functional assistance, whether this endangers the discharge of his responsibilities and whether to approach formally his immediate superior. For stating, if he does make such an approach, the nature of the service required, in order that his superior may either: (a) alter the responsibilities, (b) arrange to provide the required service, or (c) take the matter to an executive level which can arrange to provide it.
28. For obtaining from the appropriate person technical prescriptions.
29. For implementing or rejecting, clearly and with reasons, every prescription received, and for taking the full responsibility of his action.
30. For reporting to others any results of prescriptions given by them which seem likely to be of value.
31. For taking action outside the terms of policy, Standing Orders, or instructions received, in emergency when the Company's interests seem to require it, and for reporting to his immediate superior as soon as possible details of such action taken.
32. For giving his nuclear command the maximum possible amount of information and explanation, and causing that which can properly be disseminated to pass down the executive chain to his extended command.
33. For ascertaining if his instructions are being effectively carried out, and whether there are difficulties which may be slowing down their execution, or which may suggest modification.

332

34. For creating, by his own behaviour, the pattern which he desires his subordinates to follow.

35. For training, or organizing training for, such members of his nuclear command as require it, and holding them responsible for the training of their subordinates.

36. For deciding when to ascertain the facts and feelings about any matter from his own subordinates, the Company's specialist services, or outside sources of information, for the purpose of making decisions, or helping his executive superior to do so.

37. For maintaining as his nuclear command a team of individuals who are competent to discharge the responsibilities delegated to them.

38. For the establishment in his extended command of a sound executive system.

39. For the maintenance in his extended command of efficient administrative practices.

40. For the use and promotion of agreed formal joint consultation practice in his extended command for the institution or amendment of matters of policy or principle.

41. For stating differences of opinion with his immediate superior to that superior, and for encouraging his own immediate subordinates to state to him their opinions, whether or not they be in accord with his own.

42. For discussing with the individuals concerned personal conflicts between members of his nuclear command.

43. For manifesting his readiness, if it be desired, to submit any difference of opinion between himself and a member or members of his nuclear command, to his own immediate superior.

44. For protecting his subordinates from unjust treatment.

45. For maintaining as confidential differences of opinion between himself and his immediate superior, except when it is jointly and specifically agreed that they may be disclosed. For seeking to have such differences considered by a higher level of executive authority when this seems necessary for their resolution.

POLICY GOVERNING FUNCTIONAL RELATIONSHIPS

When there is a Functional Relationship between two people, the servicing or prescribing person shall carry responsibility:

46. For the quality of any prescription or service rendered, but such responsibility for quality shall be to his own executive superior only.

47. For giving prescriptions in the field of knowledge specifically assigned to him, either when it is clear to him that it is in the Company's interest to do so, or at the request of an authorized person.

48. For using his best informal endeavours to get his prescriptions accepted, but for recognizing that the authorized person requesting service carries full responsibility to his own executive superior for the results of either accepting or rejecting a prescription, and therefore must make his own decision on the matter.

49. For rendering, at the request of an authorized person, services of any type which fall within the responsibilities delegated to him by his executive

superior which he is capable of providing. Such services to be given with the minimum possible delay.

50. For using his best endeavours informally to dissuade an authorized person from seeking a service which he feels is inappropriate, but for recognizing that persons requesting service do so in order to be able to discharge their executive responsibilities, and are therefore entitled, if they insist, to receive any available service.

51. For clearly informing those who seek services which he is not capable of providing of that fact.

52. For refusing service to, or taking action designed to stop any person who is performing or is about to perform an act which is deemed to be seriously damaging or illegal.

REFERENCES

The main ideas which have been used in this work have derived from developments reported in:

BION, W. R. "Experiences in Groups", I–VI, *Human Relations*, Vols. I–III, 1948, 1949, 1950.

BION, W. R., and RICKMAN, J. "Intra-group Tensions in Therapy: Their Study as a Task of the Group". *Lancet*, Vol. II, 1943.

CURLE, C. T. W., and TRIST, E. L. "Transitional Communities and Social Re-connection". *Human Relations*, Vol. I, Nos. 1 and 3, 1947.

FREUD, S. *Group Psychology and the Analysis of the Ego*. London, International Psycho-Analytical Press, 1922.

JAQUES, ELLIOTT. "Interpretive Group Discussion as a Method of Facilitating Social Change". *Human Relations*, Vol. I, No. 4, 1948.

KLEIN, MELANIE. *Contributions to Psycho-Analysis*. London, Hogarth Press, 1948.

LEWIN, KURT. *Dynamic Theory of Personality*. New York, McGraw-Hill Book Company, 1935.

LEWIN, KURT. "Frontiers in Group Dynamics". *Human Relations*, Vol. I, Nos. 1 and 2, 1948.

MAYO, ELTON. *The Human Problems of an Industrial Civilization*. New York, Macmillan Co., 1933.

MORRIS, B. S. "Community Studies and Community Education in Relation to Social Change". *Occupation Psychology*, Vol. XXIII, No. 3, 1949.

MORENO, J. L. *Who Shall Survive*. Washington, Nervous and Mental Disease Publishing Company, 1934.

MURRAY, H. A. *Explorations in Personality*. New York, Oxford University Press, 1938.

PARSONS, T. *Essays in Sociological Theory Pure and Applied*. Illinois, Free Press, Glenn Co., 1949.

References

TRIST, E. L., and Bamforth, K. "Social and Psychological Consequences of the Longwall Method of Coal-Getting". *Human Relations*, Vol. III, No. 4, 1950.

WILSON, A. T. M. "Some Implications of Medical Practice and Social Casework for Action Research". *Journal of Social Issues*, Vol. III, No. 2, 1947.

Reports on the Glacier Project
Other articles describing the Glacier Project appear under the general series title "Studies in the Social Development of an Industrial Community (The Glacier Project)" in the Journal, *Human Relations*. Titles published to date are:

JAQUES, ELLIOTT. "Collaborative Group Methods in a Wage Negotiation Situation", (The Glacier Project—I). *Human Relations*, Vol. III, No. 3, 1950.

RICE, A. K., HILL, J. M. M., and TRIST, E. L. "The Representation of Labour Turnover as a Social Process "(The Glacier Project—II). *Human Relations*, Vol. III, No. 4, 1950.

Projected articles in this same series will include reports on: the sorting out of the problem of authority in functional relationships between managers; a follow-up study of the Service Department; a follow-up of the effects of the Works Council change; a psycho-social analysis of problems of mass production as seen in the Line Shop; and technical reports on the methods used and the theoretical consequences of the work.

INDEX

INDEX

TITLES IN THIS SERIES